Exploring Our Dreams

Exploring Our Dreams

The Science and the Potential for Self-Discovery

PAUL R. ROBBINS

Exposit

Jefferson, North Carolina

LIBRARY OF CONGRESS CATALOGUING-IN-PUBLICATION DATA

Names: Robbins, Paul R. (Paul Richard), author.
Title: Exploring our dreams : the science and the potential
for self-discovery / Paul R. Robbins.
Description: Jefferson, North Carolina : Exposit Books, 2018. |
Includes bibliographical references and index.
Identifiers: LCCN 2017061171 | ISBN 9781476672755
(softcover : acid free paper) ∞
Subjects: LCSH: Dreams.
Classification: LCC BF1078 .R63 2018 | DDC 154.6/3—dc23
LC record available at https://lccn.loc.gov/2017061171

BRITISH LIBRARY CATALOGUING DATA ARE AVAILABLE

ISBN (print) 978-1-4766-7275-5
ISBN (ebook) 978-1-4766-3180-6

Front cover image © 2018 iStock/yulkapopkova

Printed in the United States of America

Exposit is an imprint of McFarland & Company, Inc., Publishers

Exposit

*Box 611, Jefferson, North Carolina 28640
www.mcfarlandpub.com*

For Sharon—
bright, caring, and unfailingly helpful

Acknowledgments

I would like to thank Sharon Hauge for her help throughout the writing of this book. I would also like to thank my brother Larry, my niece Wendy and good friend Martha Weaver for their contributions.

This book includes material reprinted from several sources. The two primary sources are my book *The Psychology of Dreams*, published by McFarland in 1988, and a paper I wrote detailing the procedures used in developing the Dream Incident Technique. I wish to thank Sage Publications for its permission to include this material here. The original paper was "An Approach to Measuring Psychological Tensions by Means of Dream Associations" and published in *Psychological Reports* 18 (1966), 959–971.

I also included some brief quotations from Carl Gustav Jung's *Dreams*, published by Princeton University Press in 1974, and from papers appearing in the *Bulletin of the Menninger Clinic*, the *Journal of the American Psychoanalytic Association*, and the *Proceedings of the National Academy of Sciences of the United States of America*. While formal permission to include these brief quotations was not required, I wish to thank the editors for their courteous, positive responses to my inquiries. Complete citations for the book and papers are in Notes and Sources.

I included detailed descriptions of some of the research findings from the pioneering work of Calvin Hall and Robert Van de Castle, published in the *The Content Analysis of Dreams* in the 1960s by Appleton-Century-Crofts. I wanted to bring their groundbreaking research to the attention of contemporary readers.

Finally, I wish to acknowledge the work of the many researchers from psychology, psychiatry, and neuroscience whose studies are described here. Without their creative application of the methods of science to the study of dreams, this book would not exist.

Table of Contents

Introduction

Interest in dreams goes back a long way, to the dawn of written human history and probably deep into the recesses of human experience. An Egyptian papyrus dating from thousands of years ago contains more than 100 ancient dreams with interpretations of their meanings. The Bible contains fascinating stories of dreams and their interpretations and Homer included dreams in his epic tales *The Iliad* and *The Odyssey*. When I was an active dream researcher, I wrote a book, *The Psychology of Dreams*, published by McFarland, which presented a comprehensive look at what we knew about dreams at the time. Since its publication, a great deal of interesting research has been conducted on dreams and we now have more insight into both their nature and their potential use in informing us about ourselves. Looking through some of this new information, I began to think it was time to write a new book about dreams. What prompted me finally to go ahead was a telephone call from a young student in Taiwan who wanted to interview me about dreams for a class project. It was a pleasure talking with her and answering her questions. I decided it was time to go ahead and write the new book.

Let me offer a brief description of what readers will find here. The first part of the book explains what we have learned about dreams from decades of psychological research, from research carried out in sleep laboratories, and from the more recent investigations using the new technologies of neuroscience. The second part delves into the methods that people of all sorts, soothsayers, purveyors of dream dictionaries, psychoanalysts, and psychologists like myself have developed to attempt to understand what our dreams might be telling us about ourselves. The first part is grounded in the methods and research of science. The second part is more speculative with only occasional efforts to use the methods of science to demonstrate that the type of dream analysis proposed has any validity. Nonetheless, I will show how dream analysis can be more objective and how an approach to dream analysis based on scientific research can be useful in understanding some of the problems that confront us in our daily lives.

While this book may be of interest to those who teach psychology or practice psychotherapy, it is primarily intended for the general reader. I have tried not to be overly technical. The first part of the book is written in a conversational style, in a question and answer format. The second part is written more conventionally. I would like to assure the readers that the information presented in both parts is firmly based on research. My Notes and Sources clearly testify to that.

Why the question and answer format? Because it is simple and direct. I have used this format in a previous book, *Romantic Relationships*. I enjoy writing this way, readers have told me they like the conversational tone and critical reviews have been positive. I believe the topic of dreams lends itself very nicely to this straightforward approach.

A word about the questions. I have asked people online about the questions they were interested in and have included a few of these. However, most of the questions reflect the questions researchers have asked themselves when they study dreams. I have simplified their questions, eliminating technical terms as much as I can, and putting them into plain language.

A bit about my background as a dream researcher and psychotherapist: I hold a Ph.D. in psychology and I enjoyed a long career both as a researcher and as a psychotherapist. While I was on the staff of the Department of Psychiatry at George Washington University, I conducted many studies on dreams that were published in psychological and psychiatric journals. During my years as a psychotherapist, my patients often reported their dreams to me, and in the discussions that followed their dream reports, both the patients and I found the conversations useful in more clearly understanding their problems.

My earlier book, *The Psychology of Dreams*, contained a great deal of information about dreams based on the research available at the time. In this new book, I have focused heavily on recent research. I hope it will stimulate your interest in this tantalizing subject.

PART I

What Science Tells Us About Our Dreams

1

The Nature of Dreaming

What is a dream?

At first blush, this may seem an unnecessary or a silly question. We are all aware of what dreaming is, but it may be hard to define. A Supreme Court justice, once asked to define "pornography," said he didn't know how to define it precisely but he knew it when he saw it. Dreaming may be like that. Even so, when I said all of us are aware of what dreaming is, that isn't quite true, at least from what some people report. Some people assert they never dream. Studies in sleep labs indicate that at least some of these non-dreamers actually dream but simply can't recall what they dreamed about. Leaving these folks aside for the moment, dreams have a common-enough meaning that they have found their way into our poetry, music and literature. Robert Louis Stevenson began one of his poems by writing that he dreamed of forests and fields of gray-flowered grass. There have been hundreds of songs with the word "dream" in their titles, sung by such luminaries as Elvis Presley, Barbra Streisand and Lady Gaga. We cannot forget Shakespeare's famous soliloquy in *Hamlet*, "To be or not to be," where Hamlet ponders suicide and then worries about dreaming after death.

If we turn to dictionaries for help, we are likely to see definitions like "a succession of images, thoughts, emotions, and sensations that occur involuntarily while one is sleeping." Fair enough. Dreaming is an ongoing process during sleep in which we feel we are taking part in or witnessing scenes, events, and conversations that seem real, as if they were actually happening, and typically, what we are experiencing is outside our control. Really, when you think about it, isn't dreaming like watching a movie, a very personalized movie? For it is our brains that write the script, supply the actors and producer and direct the film. Guess who that good-looking, intrepid star is. Easy. Just look in the mirror.

Perhaps the most important characteristics of dreams are that they occur during sleep, that they are perceptual experiences and that they are usually involuntary. If we add to these characteristics the facts that what we see in dreams is not likely to be static, like a painting but changing from moment to moment, that what happens in our dreams is often coherent enough to make sense of, even story-like, and that our dreams may involve our psychological and behavioral repertoire—actions, thoughts, words, and emotions—we have a pretty good description of the common understanding of dreams.

There are three more things about dreams I'd like to mention that make them so interesting. First, dreams can sometimes be fanciful, improbable, or even bizarre. Here is a snippet from a short dream a friend related to me, improbable to be sure, but with a spot of delightful imagery that brought a smile to my face:

> My sister and I were shopping and there was a slot machine there that she was playing. (By the way, I dream a lot about winning money in the slot machines and I'm always so disappointed when I wake up and it's just a dream.) Anyway, she won $5,000, and dollar bills came flying out of the machine. I remember picking up a $20 bill and being so jealous and depressed that she won and not me.

As far as the *improbable*, imagine the slot machine spitting out a torrent of dollar bills!

This dream also illustrates a second and very important aspect of dreams, at least to my way of thinking. Dreams sometimes, perhaps often, address issues that we haven't fully resolved. Our lives are seldom short of problems, major and minor, that we haven't sorted out. They may be concerns about school or work, or something interpersonal, or something more deeply about ourselves. Dreams often touch on these concerns subtly or not so subtly.

Third, dreams can and usually do rapidly disappear from our memory.

For their part, scientists might call dreaming an activity of the brain, indicating which areas of the brain are mainly involved, and further specifying that such activity is most likely to occur when one is experiencing rapid eye movement (REM) sleep, but we shall get to the scientific data shortly.

Granted that dreams occur when we sleep, when we think of dreams, we think of something that happens while we sleep at night. If we fall asleep during the day, say, take a nap, can we dream then, too?

I don't remember having a dream while taking a daytime nap, but the answer to the question is clearly *yes*. People have been studied in sleep labs while taking naps during the day. When the monitoring equipment revealed that the sleepers were experiencing rapid eye movement indicating the likelihood of dreaming, the sleepers were awakened and often reported dreaming. With the relatively short duration of a daytime nap, dream recall may be higher than for nightly dreams.

What about daydreams? How do they compare with night dreams or nap dreams during the day?

Daydreams are usually under conscious, voluntary control while night dreams and nap dreams are usually not. Daydreams can detach us from our routine thoughts, at least for a while, as we set about our daily chores, and they can offer pleasant relief from the tedium of repetitive jobs. But a daydream about a pleasant day at the beach is a huge step from the experience of a dream that comes upon us in the midst of sleep in which we enter a world that we did not ask for and can find troubling. Dreams recalled after daytime naps that took place during rapid eye movement tend to be more bizarre and sensory than the typical daydream. While both true dreams and daydreams can tell us things about ourselves, there is an important difference in that the onset of a true dream is usually involuntary and may tell us things about ourselves that we may not, in the light of day, think about or want to think about.

Many daydreams in some way or another may concern problems, perhaps a romantic issue or an issue in the home or on the job. In one study, women reported they did more daydreaming than men did. The young men in the study reported the highest number of erotic daydreams.

Why am I not surprised? You mentioned REM sleep. It sounds like a benchmark for dreaming. Tell us more.

Rapid eye movement is an important signal that dreaming is taking place. Before the discovery of this fact, researchers had to depend entirely on the dreamer's recall of his or her dream. Theorists like Freud or Jung had to depend entirely on what their patients reported to them after some time had elapsed, usually hours, days or even weeks after the dreams had happened. As many of us can testify, the memory of dreams is often fleeting, disappearing within moments of waking, so what really transpired in the dream and what was reported subsequently might be quite different. The discovery of the connection between rapid eye movement and dreaming made it possible to obtain more accurate descriptions of dreams. This is a significant advantage for contemporary theorizing about dreams.

REM sleep is a part of our normal sleep, perhaps 20 percent of our total sleep. Our eyes move more rapidly than when we are in the other 80 percent of our sleep, NREM (non-rapid eye movement) sleep.

Many actions of our brains and bodies are altered during REM sleep. Here are some of them.

Bursts of electrical activity fire up from the brain stem.

Low-voltage de-synchronized brain waves are detected.

The limbic system of the brain, which is involved in emotion, motivation and memory, is more active. Included in this activation is a small area of the brain called the *amygdala*, which plays an important role in such emotions as anger.

Our heart rates and respirations are more rapid.

While our eyes are darting about, for most of us, our bodies remain quiet. Our eyes seem to be looking, even searching, while our bodies are inert.

If we are abruptly awakened during REM sleep, there is about a four out of five chance we will report dreaming. The longer the period of REM sleep lasts, the longer the subject's report of the dream is likely to be. The measure used here is simplicity itself—the number of words in the dream report.

It has been suggested that the fact that our bodies remain quiet while we are in REM sleep and probably dreaming has adaptive significance. Imagine having a nightmare about being chased by a monster (you spent too much time during the evening watching a horror movie), and instead of lying still during your dream, you were able to react by leaping out of bed and running at full speed. You might end up fracturing your hip.

The discovery of REM sleep and its connection to dreaming seems like a game changer, a real breakthrough. Was this discovery part of a carefully designed research program in which researchers were looking for this tie-in, or was it more of an accident like the discovery of penicillin?

From what I have read, it was not something the researchers were looking for. The discovery of rapid eye movement and its connection to dreaming is an interesting story and it involves some interesting people. The discovery of REM sleep would have been difficult if not impossible without the advent of the sleep laboratory. Today, we have modern sleep labs around the country that help physicians better understand and treat sleep disorders that many people experience. These include problems like insomnia, sleep apnea and sleepwalking. The American Academy of Sleep Medicine has accredited more than 2,500 such sleep centers in the United States. These sleep centers use comfortable rooms that remind one of a hotel room, employ video cameras so that the patient may be observed throughout the night, and use modern technology to measure brain waves, eye movement, muscle activity, breathing, heart action, and other measures of reactivity. Sensors placed on the subject's scalp, temple, chest and legs with many wires extending outward ultimately come together in a box that in turn is connected to a computer. The technician who examines the data will see many rows of wavy lines, offering a multifaceted look at what is happening to the patient during the night.

The situation was entirely different around 1950 when Nathaniel Kleitman set about establishing a laboratory to study sleep at the University of Chicago. Kleitman, born in Russia during the days of the last czar, fled his home country to avoid the pogroms orchestrated to harass and terrify its Jewish inhabitants. He went to Palestine, lived there for a while, and then immigrated to the United States.

Financially strapped, he put himself through the City College of New York and went to the University of Chicago where he earned a Ph.D. in physiology, doing his doctoral dissertation on sleep.

Centers that might warrant the title of a laboratory to study sleep were rare or nonexistent when Kleitman began to develop his laboratory in Chicago. His work was truly pioneering. By today's standards, Kleitman's laboratory would seem very primitive but it was a groundbreaking effort to begin studying sleep seriously.

The man at the controls of Kleitman's laboratory the night REM sleep was discovered was Eugene Aserinsky, one of Kleitman's graduate students, who was working toward a Ph.D. Aserinsky, also the son of a Russian immigrant, took a position in Kleitman's laboratory and began to monitor the sleep of children and adults. One of his early subjects was his own son.

A third pioneer in sleep and dream research, William Dement, also a student of Kleitman's, wrote about Aserinsky's discovery of REM sleep. He related that Aserinsky was to monitor the eye movements of a subject at different stages of sleep. While the equipment in the laboratory used for this task was primitive from today's perspective, it did provide a continuous recording of the subject's brain waves and eye movement as well as other data as the subject slept. As the night progressed, Aserinsky looked at the recordings of wavy lines on chart paper detailing the various measures. He observed that there were periods of time when the subject was clearly sleeping, but his eyes were not resting. Rather, they were moving about as if they were looking at something.

These eye movements in a sleeping person were unexpected and surprising. Try to imagine what it would have been like to be in Aserinsky's shoes that night. He could see a person clearly sleeping, eyes closed, his body resting, and there on the chart paper he was studying was unmistakable evidence that, at times, his eyes were darting about, as if he were awake, looking at something, perhaps something moving, perhaps viewing different objects. Whatever it was, it did not appear to be a fixed object, but something seemingly changing. The dramatic contrast between the sleeping body and the rapidly moving eyes would be perplexing to say the least.

As trained scientists, the researchers in the lab tried to replicate the finding. It was no fluke. Further research confirmed the reality of REM. It did not take long to tie in eye movement with dreaming. A new paradigm for the study of dreams came into being.

So REM is an important signal for dreaming. What would happen if we diminished REM sleep? How would that affect our dreaming?

There have been studies that have attempted to diminish REM sleep and then look for consequences. The drug clomipramine, an antidepressant, has been used to partially suppress REM sleep (other antidepressants may also have this effect). Having said this, the typical procedure has been to monitor the subject's eye movements and brain waves, and as soon as REM sleep is about to commence, to awaken the subject. Here is an example of how this procedure was carried out in a recent experiment: the subject was asked not to consume drinks containing caffeine or alcohol and to avoid napping during the days preceding the study night. After the subject went to sleep in the laboratory, the research plan called for waking the subject every time the instruments (e.g., the EEG) showed the impending onset of REM sleep. When the signals for REM sleep were observed, a member of the research team entered the room where the subject was sleeping. Leaving the lights off, the researcher switched on a recording that played the name of the subject over and over, gradually increasing the volume until the subject woke up. The subject was kept awake by conversation for two minutes to avoid any relapse into REM sleep. This interruption of the normal sleep cycle would greatly reduce the amount of REM sleep.

One of the major findings from studies interrupting REM sleep is that subjects are more irritable and anxious the next day. They are less relaxed and more edgy. Their ability to get along with people may be diminished and they could be more confused and even suspicious. The lack of REM sleep seems to have an effect on one's emotional functioning. Some parallel findings have been reported using animals as subjects.

Another interesting effect of REM deprivation is what has been termed REM rebound. When REM sleep is artificially diminished, the next time one is going to sleep, there will be a tendency for the person to experience quicker and longer periods of REM sleep. REM rebound suggests that there is a biological need for REM sleep and that depriving one of it will increase the need for it. As a rough analogy, think of fasting all day, then coming home and overeating.

So, REM sleep may have some effect on the way we feel and behave. You said that during REM sleep, it appears that while one's eyes are darting about, one's body is typically quiet. However, sometimes we hear about people thrashing about in their sleep and they report dreaming. Is this really the case?

Yes. It is not typical but some people do thrash about, sometimes violently, during REM sleep. To use technical language, what we see is the partial or complete loss of the normal muscle *atonia* that is a hallmark of REM sleep. This thrashing about sometimes can be violent, posing a risk of injury not only to the dreamer, but also to the person he or she sleeps with. A study carried out at the Mayo Clinic noted that 32 percent of the people studied with this problem had injured themselves and 64 percent had injured their spouses. In two of the cases in the Mayo study report, the people had been involved in scary situations. In REM dream states, one person had fired a gun and a second had tried to set fire to the bed.

As the study carried out by the Mayo Clinic shows, lashing out during dreams can be a very serious problem not only for the person who does it, but also for the person sleeping with the dreamer. What takes place can be confusing, even incomprehensible, for both the dreamer and his sleeping partner, can result in serious injuries, and threaten, if not destroy, the relationship. Here is a case reported from a university medical school that graphically illustrates these consequences.

The dreamer was a 28-year-old man who since the age of 20 had shown problems of sleepiness, cataplexy (a sudden loss of muscle strength) and episodes of screaming and kicking during sleep. When he was 23 years old, he married a younger woman who reported that during the night he had talked, screamed, and occasionally kicked and slapped her while he was asleep. The situation came to a head early one morning, about four a.m., when he suddenly assaulted her.

He violently punched her and then remarkably dropped off to sleep immediately. She left the bed, went into the other room and the following morning went to the emergency room. The physician who treated her for a hematoma notified the hospital police but the woman refused to press charges. The husband was both mortified and astonished, having no recollection of assaulting his wife. After another incident of being hurt by her husband, she began to sleep in a separate room, eventually separated from and divorced him, and reported him to the police.

It is interesting that when the woman was interviewed she wasn't convinced that what her husband had done was unintentional, despite her clearly saying that her husband was *sleeping* during the violent episodes. The patient was treated medically and eventually the symptoms were brought under better control. He still faced both civil and criminal trials.

Researchers have observed that during REM sleep, people with this problem may be flailing their arms, punching, kicking, and very often talking as they dream. Some people (a small minority) will leave the bed and sleepwalk. Observers have noted that when the sleeper talks, his voice is louder and may be better articulated than when he is awake. Interestingly, the dreamer may act out behaviors that are reminiscent of the individual's daytime life. Retired patients have been observed practicing their former work habits. Sentences uttered by people going through this experience are often fluent and accompanied with appropriate gestures. The behavior during these more quiet episodes often reflects what the dreamer is accustomed to doing when awake.

When the patients have been awakened after violent episodes and asked about the dreams they were having, very often these dreams are aggressive, matching the violent thrashing of their bodies. The aggressive acts in the dreams are usually not something the dreamer initiates but are responses to some perceived attack, often by humans, sometimes by animals.

Rapid eye movement continues along with the dreams and body movements. Researchers have wondered whether the direction of the eye movement tends to correspond with the direction of the body movements. If the hand of the sleeping individual reaches out, as if trying to grab something, do his eyes move in the same direction? Interestingly, research suggests that most of the time that is the case. When rapid eye movement is present, it appears to imitate the action of the dream scene.

These body motions during REM sleep are now considered a sleep disorder with the name *sleep behavior disorder*, abbreviated as *RBD*. One of the most important findings about the sleep disorder is that its occurrence is associated with neurological disorders, particularly Parkinson's disease. In the Mayo study, 57 percent of the people studied had neurological problems. Many people who experience RBD will later develop Parkinson's disease.

Is RBD treatable?

Yes. The drug Clonazepam is usually effective in relieving symptoms for most people. In the case reported above, the drug sodium oxybate was found to be useful.

I'm glad that RBD is a rare, rather than a typical, problem. It is reassuring to know that during REM sleep our bodies are very still, not thrashing about. Returning to normal dreaming, we know we dream at different times during the night. Are there a number of different REM periods during the night?

Yes. You probably have heard there is a regular sleep cycle during the night that takes about 90 minutes to complete. A sleep cycle begins with very light sleep—the kind you experience as you drift into sleep. It is a short period of time in which you can be easily awakened. It is appropriately called *stage one*. In *stage two*, which lasts much longer, sleep grows deeper. In *stage three*, sleep becomes deeper still: stage three may be referred to as *delta sleep* or *slow wave sleep*. Stage three does not last as long as stage two but it is normally restful and restorative. Paradoxically, it is during this time of deep sleep that we may observe unusual behaviors like sleepwalking. Some references list a *stage four*, a time of deep sleep with more delta waves, while other references combine Stages three and four into a single entity. In any event, after deep sleep, there is a transitory return to stage two and then the onset of the first occurrence of REM sleep. REM sleep usually occurs about 90 minutes or so after initially falling asleep. The initial period will be relatively short, perhaps lasting about 10 minutes. As sleep progresses through the night, the sleep cycle repeats itself, and interestingly for dreaming, the length of time of REM periods tends to increase. Therefore, the length of your dreams may become much longer as you sleep through the night. As dawn approaches and the alarm clock has yet to ring, there is much more time for a complex dream. Indeed, research has confirmed that in a late period of REM sleep (the fourth time around), dreams were more organized, coherent and story-like than they were in an earlier period used for comparison (the second time around).

As to the number of REM periods you may experience during the night, you can get a rough estimate from simple arithmetic. If you sleep seven to eight hours a night that would be something like 420 to 480 minutes. Divide either figure by 90 minutes, which is about the typical length of a sleep cycle, and the result is four or five sleep cycles. As each sleep cycle contains an REM period, you can see that you might expect to have something like four or five REM periods during the night, and you could have as many dreams as that during the night. You probably won't remember anything like that many dreams, but that is what is reported from studies in sleep laboratories where subjects are awakened during REM periods.

One fact here may be of particular interest to parents of infant children. Newborns spend a lot of time in REM sleep. If they are in fact dreaming, what could they be dreaming about? Dreams tend to include autobiographical memories and infants haven't had time to accumulate much to write home about. I wouldn't hazard a guess as to what their dreams are like.

If you remember all of your dreams during the night, can the dreams form a connected storyline, something like the chapters of a novel?

It can happen, but it is probably an unusual event. In an early study in which two subjects slept in a sleep laboratory over two weeks, the researchers noted a single instance where there was a clear-cut sequence of dreams. During that night, one of the subjects reported four dreams with a definite storyline. In the first dream, the subject was in a garage cutting up some wood to put on the truck to take away. In the second dream, he was cutting up a door getting it ready to add to the pile of wood. In dream three, he had finished the work in the garage and then locked it. In the final dream, the garage was all cleaned out.

It is, of course, possible that a series of dreams through the night may have an underlying unifying theme, but a continuous evolving storyline like the dreams above are probably unusual. Dreams occurring during the night generally seem independent of one another.

Can we have dreams in the absence of REM sleep, and if so, what are these dreams like?

Some researchers prefer to use the term *NREM mentation* rather than NREM dreams. Regardless of what it is called, NREM mentation is likely to be different from the dreams that occur during our REM sleep. REM dreams tend to be longer, more vivid and intense, and involve stronger emotional content then NREM mentation. If there are other people involved in the dream, the dreamer is more likely to act aggressively during REM sleep. During NREM periods, the dreamer is more likely to demonstrate friendly behavior. On a much-diminished scale, this difference recalls Robert Louis Stevenson's story of Dr. Jekyll and Mr. Hyde.

Typically, NREM mentation is more thought-like and more apt to deal with one's everyday concerns. Compared to REM sleep, the mentation of NREM periods is more mundane. However, the differences between the content of REM dreams and NREM mentation become less clear in the latter stages of sleep. In the longer dreams that may appear at this time, NREM mentation more closely resembles what we usually report as dreams.

David Foulkes, a very productive dream researcher over the years, has provided some examples of the difference between dreams from REM sleep and NREM sleep. The first two examples are from an NREM sleep, and the third example is from an REM sleep:

1. He asked a person for a hammer so he could repair something in his apartment.
2. He was thinking about a point brought up in his tax class

that he would have to provide more than half of a person's support to claim the person as a dependent.

3. He received a phone call in the middle of the night from a girl saying she was from the University of Chicago. She told him it was time for his 35-day evaluation. He chided her for calling so late at night. She said it was the only time she could reach him.

You can see that the third excerpt from REM sleep has much more of the quality of what we expect in dreams.

Dreams reported after waking in the sleep laboratory during REM sleep seem to be the most reliable way we have to obtain accurate dream reports. That would seem the gold standard for obtaining an accurate report of a dream. How do these dream reports compare with the dreams we remember after waking from a night's sleep at home?

Dream reports obtained in the sleep laboratory are certainly the gold standard for accuracy. The participants are awakened during or right after dreaming and their memories are fresh. Still, there is an issue here in that the equipment used to inform the technician that an REM period is occurring sometimes finds its way into the dream. This effect is analogous to the *observer effect* noted in physics where the procedures used in obtaining the data may disturb the phenomenon under observation. Home dream reports, of course, are subject to fading, as after one awakens and the day goes on, one's memory of the content of a dream dissipates. Often the forgetting happens quickly, and, as researchers sometimes do, asking people about whether they ever dreamed about something specific in the past can taint the recollection.

It's difficult to make exact comparisons between dreams reported in the laboratory and dreams reported at home because we don't know what stage of sleep the sleeper was in when the home dream occurred. We could be comparing a dream that occurred in NREM sleep near daybreak at home with a short dream that occurred in an early REM period in the sleep laboratory. Also, the number of cases for which we have good REM data is relatively small, so our estimates of what are typical REM sleep dream reports is not all that reliable. Having said that, researchers have studied the content of home dreams and compared them with dreams reported after being awakened in the laboratory using all sorts of indicators. For example, how long the dreams were, what the dreams were about and what kind of emotions were expressed. Dream researcher Bill Domhoff took a careful look at these studies and concluded that the differences between dreams reported in the laboratory and dreams reported upon waking are not all that different.

One of the most pronounced differences was that dreams reported at home seemed to have more aggression than dreams reported in the laboratory. We really don't know why this is the case, but it is certainly a provocative finding.

REM sleep and NREM sleep typically produce different kinds of dream content. Do these differences reflect the involvement of different areas of the brain?

This is a question that we can now begin to look at seriously because of the development of increasingly sophisticated methods of brain imaging. These techniques permit researchers to look at the structures of various parts of the brain and to assess which areas are active during our various experiences. Such imaging allows us to visualize which areas of the brain are active during both our waking and sleeping experiences. Two of the brain imaging techniques that have been used to study sleep—including both REM sleep and NREM sleep—are positron emission tomography (PET) and functional magnetic resonance imaging (fMRI).

A few words about each: positron emission tomography (PET) assesses emissions from chemicals that have been rendered radioactive and injected into a person's bloodstream. These radioactive chemicals will work their way to the brain where they are detected by the PET scanner. With the aid of a computer, images are created from these data that show which areas of the brain are most active when a person is presented with different tasks when awake or what is happening within the person's brain during the different stages of sleep.

2

Dreaming and the Brain

Most of us are familiar with traditional MRIs. We receive them in hospitals or imaging centers and they allow physicians to more closely examine the structures of different areas of the body, particularly when we are experiencing pain or discomfort from an area of the body. Functional magnetic resonance imaging (fMRI) may assess how a structure in the brain is acting rather than just what it looks like. Images of blood flow allow researchers to obtain measures of oxygen in the blood that indicate which brain structures are activated.

As is true for much of current scientific research, research on the areas of the brain that are involved in REM sleep and NREM sleep is often collaborative, and the publication and interpretation of what these results mean often involves collaborative work of scientists from different universities and research settings. Some of the most recent reports come from the collaborative work and thinking of scientists from several European countries such as Belgium, Austria, and Switzerland. Drawing on one of their most recent papers, I would like to highlight some of their findings regarding the activation and suppression of specific areas of the brain during REM sleep and NREM sleep.

Briefly, REM sleep is associated with an activation of the pons (a part of the brain stem that relays messages), thalamus (also involved in relaying messages), limbic areas (heavily involved in our emotions) and the temporal occipital cortices (involved in sensory activities). There is an accompanying decrease in the activity of the prefrontal areas (which deal with executive functions of the brain, helping us guide our choices and behavior).

NREM sleep is associated with decreases in brain activity in the brain stem, the thalamus, and in several cortical areas.

As a generalization, it is fair to say that during REM sleep, the brain seems as active as it is during waking hours with some areas of brain showing even greater activity than when the person is awake. In contrast, during NREM sleep, the activity of the brain is markedly reduced. In their review of research, Dang-Vu and his colleagues noted that researchers have found a decrease in brain activity during NREM sleep when compared with wakefulness. They noted that this decrease was estimated to be about 40 percent during slow wave sleep. During NREM sleep, the brain appears to be more quiescent. Not completely, of course, as we know there are mental activities occurring during NREM sleep, but the activity of the brain is clearly reduced, and interestingly, the intensity of the action in our dreams is more subdued. There is clearly more friendliness and less aggression in our dreams.

This, of course, is a quick summary. In this book, I have tried to avoid becoming too technical, but in the interest of completeness, I would like to list in more detail the areas of the brain that become more active and less active during both REM and NREM sleep. Some of the terms used in the geography of the brain may sound very intimidating and I am not urging readers to master them. Consider the first structure mentioned, the pontine tegmentum. My first reaction to the term was it sounds like a bridge across the River Arno in the city of Florence, Italy. That shows you where I am, so don't take the next page or two as a challenge. While the listing could serve as a starting point for those who are interested in learning more about the workings of the brain, such study is not at all necessary to carry the points I want to make. Personally, I find it more interesting to pay attention to what these structures do rather than to their names.

I have two major reasons for listing the specifics. The first is to show clearly that when we dream, it is not the action of a single area of the brain. Dreaming is a very complex set of actions that are taking place in several areas of the brain. The second reason is to demonstrate that the activity of the brain during REM sleep is clearly different from NREM sleep. The authors of the paper I just cited emphasized this point. They concluded that the patterns of brain activity during REM sleep are drastically different from NREM sleep.

Here is a summary of brain activation and deactivation by researchers who have been studying this question for a number of years. Once again, our primary source is the article written by Thanh Dang-Vu, et al., published in *Sleep*, in December 2010.

These brain structures *enhance* their activity during *REM* sleep when compared to waking:

- the pontine tegmetum, located in the brain stem, is one of the two parts of the pons and controls stages of sleep and levels of arousal and vigilance;
- the thalamus, a symmetrical structure in two halves located between the cerebral cortex and the midbrain. It has been described as a kind of information hub that relays signals from the subcortical areas to the cerebral cortex. It plays a role in regulating sleep and wakefulness;
- the basal forebrain, a group of structures that help produce the neurotransmitter acetylcholine. Acetylcholine promotes wakefulness and REM sleep;
- the amygdala, part of the limbic system. It consists of two almond-shaped masses—a left and a right—located in the temporal lobe of the brain, and it plays an important role in our emotional reactions such as anger and fear;
- the hippocampus, another part of the limbic system, resembling a seahorse in appearance. Once again, there are a pair of these structures, one located in the left part of the brain, the other in the right part. The hippocampi play a role in the formation and consolidation of memory;
- the anterior cingulate cortex, which plays a role in regulating heart rate and blood pressure and in processing emotions; and
- the temporo-occipital areas, which assemble and make sense of incoming visual and auditory information and transmit it to other parts of the brain.

These structures *decrease* their activity during REM:

- the dorsolateral prefrontal cortex, a recently evolved area of the brain that doesn't fully develop until adulthood. It plays an important role in memory, although its precise function is not completely understood. One could think of it as a part of the executive functioning of the brain, involved in judgment and decision;
- the posterior cingulate gyrus, whose function has been described as still "mysterious." It is a highly connected area of the brain believed to be involved with the retrieval of memories and has been hypothesized to play a role in altering one's behavior in dealing with unexpected changes;
- the precuneus, which is involved in memory, and interestingly, in memories related to the self; and
- the inferior parietal cortex, a structure not far from the visual areas of the brain that helps interpret sensory information. It also deals with language.

You can see why I stressed that dreaming is very complex, as during REM sleep, a lot of areas of the brain are revved up or quieted down.

Let's discuss NREM dreams. During *NREM* sleep, there does not appear to be an overall activation in the various areas of the brain. While there may be fluctuations within areas of the brain, the overall picture is one of reduced activity. Specifically, reductions of activity occur in the following:

- the brain stem, which runs continuously from the spinal cord to the brain and conducts information from the body to the brain as well as the reverse;
- the thalamus, which, as described previously, relays signals;
- the basal ganglia, located at the base of the forebrain, which connects the brain stem with the thalamus and cerebral cortex;
- the basal forebrain, important in the production of acetylcholine which promotes wakefulness and REM sleep;
- the prefrontal cortex, the multilayered part of the brain that allows us to perceive and understand our environment, think, plan, use language, and react and move. I think of it as the brainy part of the brain;

- the anterior cingulate cortex, which plays a role in regulating heart rate and in processing emotions; and
- the precuneus, which, as described previously, is involved with memory, particularly memories related to the self.

Here I must introduce a caveat. Most of the structures in the brain that I have mentioned have more than one function. Some have quite a few functions. Most of my descriptions of the functions of the brain could have easily been preceded by the phrase "among other things." If I had listed all of the known or hypothesized functions of the areas of the brain that have been highlighted as activated or deactivated, this book would be an unreadable muddle, so I have selected one or at most a few of the functions I found most interesting when I think about dreams. When one picks and chooses, there is always the possibility of introducing a bias.

Is there a pattern discernible in the activity of the brain during REM sleep? Something that seems to match the kind of dream content we see in REM sleep? This is a hugely important question, and in describing their research findings, the neuroscientists have been appropriately cautious in linking selective brain activations to dream content. For example, in a paper carrying out meta-analysis on research using PET scans, P. Maquet talked about the findings as offering a plausible explanation for some of the features we obtain in dream reports. In other words, our current understanding of this linkage is far from complete and our inferences tentative.

I have not yet seen a claim that we can state with precision what our dreams will be like based on our knowledge of what areas of the brain are activated or deactivated. Given what we know now and what we hope to find out, there will be attempts to make general links between brain activity and some aspects of dreaming. This is certainly worthwhile. Science thrives on hypotheses. However, the task of linking brain activity and dream content may not result in clear-cut correspondences, and if this proves to be the case, the theorists who plow these furrows will have to guard against any tendency to see what they want to see as they seek to match the complex activities of the brain and the variable, sometimes chaotic scenes we see in dreams.

Seeing what one hopes to see is a common failing in everyday life. It occurs in scientific investigation as well. One of the best examples of a scientist seeing what he wants to see occurred when astronomer Percival Lowell, while looking at Mars through his large refracting telescope in his observatory in Arizona in the early part of the 20th century, noticed some lines on the planet's surface which he thought might be canals and could indicate the presence of an advanced civilization. He theorized that Martians had built the canals to obtain water from the polar caps in an effort to survive in an otherwise dying planet. His speculations turned out to be more fanciful than anything. His conclusions led to some first-rate science fiction stories, but so far, that's about it. No green men with ray guns or light sabers have so far been uncovered.

One of the great virtues of science is that it is self-correcting. Errors in collecting data, faulty interpretations of the meaning of data, and even occasional fraud are discovered and yield to views that are more correct. The big problems in seeing what you want to see are unlikely to come from the scientific study of dreams but in the attempts to understand the meaning of dreams. We shall see much too much of this in Part II.

With this caution about the interpretation of data, let's take an intuitive look at the findings presented so far, and suggest two or three notions which seem *plausible* in linking brain activity and REM dreams. Beyond this, I will leave the attempts at possible linkages to the neuroscientists who are far more capable than I am.

Link one: there seems to be no across-the-board activation of brain areas when we consider NREM sleep. The overall pattern seems to be decreased brain activity. What is the closest analogy that we can think of when considering our dreams? It seems to me it may be watching a movie, admittedly a sometimes confusing, hard to follow movie, but nonetheless a movie. Pushing this idea a little further, compared to the dreams of REM sleep, the dreams of NREM sleep look like a low-budget release, a rather unexciting script, not particularly dramatic, short on special effects, a product that would do poorly at the box office. The analogy doesn't go so far as to compare REM dreams to *Star Wars* and NREM dreams to a boring documentary, but as a general rule, NREM dreams often are on the dull side and have received the somewhat disparaging label "mentation" rather than dreams. Is the decreased brain activity during NREM dreams responsible in any way for this uninspiring product? Probably. If it were the other way around, it would be troubling.

Link two: the activation of the *amygdala* (a center for emotions in the brain) during REM sleep seems to run parallel to the frequent presence of anger and fear in dreams. So far, so good. That seems to make sense, and there's more. You will remember that there are two amygdalae, a left one and a right one. Using fMRI, researchers have measured the volume and the diffuseness of these two amygdalae. Volume, of course, is a measure of size, and diffuseness is a measure of structural integrity; the more diffuse, the less solidly the structure seems put together. Researchers have correlated these two measures of the amygdalae with the kinds of dreams that people report. One would think that a structurally sound amygdala would be associated with more emotionality in dreams than one which was not as well put together, and that is indeed the case. A more diffuse left amygdala was associated with lower scores in the subjects' emotional load in dreams. Something I find very interesting is bizarreness in dream reports was more likely to occur when the left amygdala volume was low and the right amygdala was diffuse. It appears that when these emotional centers in the brain are smaller and less structurally sound, the dream seems more likely to deviate from reality.

Link three: the actions of some of the areas of the brain during REM sleep, which stimulate wakefulness and vigilance, seem consistent with an active dreaming state.

The three examples sound plausible, but there is a lot more activation and deactivation going on in the brain that you didn't consider. Do you think it likely that all of this can be tied in neatly with the content of our dreams?

As I said, I would like to defer to the neuroscientists who are doing the research and can most competently speak about it, but as a dream researcher, I'm skeptical that they will find a tidy and complete package that will fully explain the many facets of dream content. Our bodies don't always work in a perfectly coordinated fashion, and I suspect that our brains don't either. It may be that's why our dreams can sometimes be bizarre.

Let's remove our scientific white coats and consider this question on a more philosophical level, as a thought problem, one in which we allow our imagination free reign to speculate and think in a more wide-ranging way, one in which we are held less accountable by those troubling things—facts.

Think again of the product of our sleeping brain, the dream. At times our dreams can be incoherent, jumbled, not making much sense, even bizarre. Should we really expect that the elements of our sleeping brain should be performing as a well-oiled machine with all of the elements working flawlessly with one another? Maybe we should expect a less organized response.

Let's push this idea a little further. We know that the human brain is a fantastic instrument, one that figured out the laws of gravity and the theory of relativity, unlocked the secrets of the atom and wrote the plays of Shakespeare, but even the centers of our waking brain do not always work optimally together. Let's take a common-place example. The brain has both emotional centers that can produce reactions like anxiety and fear and centers that allow us to solve complicated problems. From an evolutionary standpoint, these centers and others working in tandem must have had enormous survival value for our ancestors who lived in a world that was both challenging and dangerous. Even in our waking lives the combination of emotions like anxiety, which can be highly motivating to get things done, and our problem-solving capabilities, which permit us to accomplish what we need to do, can be valuable in our schoolwork and jobs, but we all know that the output of these brain centers can sometimes get out of balance; an excessive amount of anxiety can interfere with our ability to pursue a goal with our best efforts. Think for a moment of making a speech when anxiety produces stage fright, or taking a math exam in which our thinking processes may become muddled by anxiety or even being one's best self on a first date if one is beset by social anxiety. If what we hope to accomplish can be diminished during our waking lives, when the output of our brain centers may be only slightly out of kilter, think of the images and storylines that might come during our sleep when our judgment in the decision-making parts of the brain are diminished, our emotions are activated and the activities of some of the other parts of the brain may not even be on the same page. The product? Our dreams during REM sleep may become confusing, incoherent, and even bizarre.

So much for our thought problem. Let's put an end to our speculations and fanciful analogies, put on our white coats once again, return to science, and await what the researchers will uncover.

Some people say they clearly remember their dreams. Other people say that they hardly ever remember their dreams. What are some of the things we have learned about dream recall?

The link between REM sleep and dreaming suggests that most people dream even if they say they don't recall their dreams. Even for most of us who know that we dream and recall some of our dreams, dream recall is far from certain. Sometimes we remember our dreams very easily upon waking, but even in a few minutes, perhaps only seconds, the details of the dream may prove to be fleeting. Other times we may have the feeling we dreamed about something but can't remember anything even if we try very hard. Have you ever had this experience? You know you have dreamed, but every trace of it has disappeared from your memory. Some researchers have referred to the recognition that we have dreamed but can't remember anything about the dream as contentless dreams. There has been some research on contentless dreams and we shall say a word or two about it as we discuss dream recall.

There are a lot of things to consider when we think about dream recall, but since we have just been talking about the brain, why don't we begin there?

Research on brain activity and dream recall is still in its early stages. However, there is evidence from several lines of investigation that suggest that the neural physiologic mechanisms that underlie memory for dreams are similar to those that underlie similar types of waking memories. For example, studies using the electroencephalogram (EEG) indicate that the brain waves associated with dream recall are the same involved in recall of episodic memories during our waking hours.

3

Dream Recall, Dreaming in Color and More

Having said that, wouldn't someone with better general memory be more likely to recall dreams?

Memory is complex. There is short-term memory, involved in absorbing new information, and there is the long-term memory, stored in your brain. There are memories of the events in your life, detailed narratives that you might recount to a good friend. There are facts and names and events that you may have learned in a high school history class that may come back only with prompting. There is motor memory (also called *muscle memory*) which seems to last a lifetime. Think, for example, of sitting on a bicycle and wanting to ride it. If you did it as a child, chances are you can still do it.

Intuitively, it would make sense that a good memory for what has been happening to you in your life, your own story as you construct it, would be a good predictor for dream recall, whereas trying to recall the date when William the Conqueror won the battle of Hastings and became king of England might be less relevant.

So the memories of your life experiences are the stuff that dreams are largely made of. Are these memories more or less recent or could they go far back into early childhood?

I read many dream reports of college students in carrying out my own research, and while I made no tabulation, my impression is that the memories that gave rise to dreams tended to be more recent than very early. In my research using the *Dream Incident Technique*, which looks at the real-life experiences people bring up when talking about their dreams, certainly most of the experiences were recent.

Having said that, from time to time people tell me about having dreams that take them back many years. Often these dreams were about situations that were important to them and/or about people they had close relationships with and often the dreams elicited emotions ranging from positive feelings to a plethora of negative feelings such as regret, anger, and ambivalence (a mix of positive and negative feelings). I suspect that negative feelings are more likely to arise when dreamers reflect on situations that might have not been resolved in a satisfying way. There are times when all of us wish that we could have done something better, or that things might have been different: romantic relationships, family relationships, work relationships, you name it.

So, dreams can sometimes take us back in time. Recently, I talked to a retired television producer who told me that she had recurrent dreams about being once again in the pressure cooker of having to turn out enough episodes of a television series to fill the season. She stopped that line of work many years ago.

What about our very early memories? Early memories can be a fascinating subject in themselves as well as in their possible role in dreams. Early memories are of particular interest to those influenced by psychoanalytic theory. Freud wrote a paper about what he called *screen memories.* These are memories from early childhood that are not really accurate but are substitutions for more meaningful early events.

The followers of Alfred Adler, best known for his development of the concept of the inferiority complex, postulated that recall for dreams and recall for early memories were both expressions of one's lifestyle and therefore might have some commonalities. I found this idea intriguing and Roland Tanck and I carried out some research to see whether the two types of memories might have a positive association. We asked university students to record their dreams over a period of time which gave us a measure of dream recall and we also asked them to write down their very early memories: anything they could remember during the first three years of life and anything they could remember from years four through six.

Let's pause for a moment and ask you, do you have any memories for the first three years of life? I can recall only one, a birthday party, my third, thrown by an aunt who was staying with us.

Tanck and I found that less than half of our university subjects were able to recall any memories during the first three years. In contrast, 90 percent of the students were able to recall memories between the years four through six. We did not find any relationship between dream recall and memories for the first three years, but we did find a significant relationship between current dream recall and memories for years four through six. So, there was some support for Adler's view. Incidentally, in his paper on screen memories, Freud cited research of others who stated they found evidence of early memories dating back to the first year of life. Upon reading this, the skepticism meter in my brain started flashing and rose off the charts.

Does mood influence dream recall? If you feel particularly happy or particularly unhappy, are you more likely to remember your dreams?

There are many shades of moods. You might feel elated, sad, angry, anxious, sexually aroused, or combinations of these and other moods. In terms of moods influencing your recall of dreams, I would like to discuss the effects of an unhappy mood state, a feeling of being down, perhaps a little depressed, perhaps a little troubled, and feeling a little lonely. We are not talking here about anything resembling clinical depression or any form of mental illness, just the kinds of ups and downs that people experience as part of daily life.

During my years of being a researcher, I developed what I called a psychological diary asking people to report about some of the experiences and feelings they had during the day. I included a number of questions that statistical analysis had linked together to indicate the experience of having had a lousy day. To give you the flavor of these questions, I asked them to report whether they had felt depressed during the day, whether they had felt lonely or isolated, whether they had things on their mind like unresolved problems, and whether they had experienced feelings of frustration or defeat. You can see right away that this kind of day with a resulting dysphoric mood is something almost everyone would experience at one time or another. If you never had such days, you would be a very unusual person.

In one of our studies using the psychological diary, we included extra pages for our research subjects—once again psychology under-graduate students—to record any dreams that they might have had during the night that they had responded to these questions. The students were more likely to report that they had a dream during the night and were able to describe it upon waking in the morning if their responses to the questions indicated they had been in a lousy mood during the day. Therefore, these results suggest that if you have been in a dysphoric mood during the day or in the evening you are likely to remember that you had a dream the following morning and would be able to describe it as well.

Remember when I spoke briefly about contentless dreams? I never liked the term because it suggests that you have a dream with-out anything in it, something like watching a television set with a blank picture, which to me sounds nonsensical. What it actually refers to is that if you searched your memory in the morning, you might say something like, "Yes, I had a dream, but darned if I can pull back a thing from it." Freud developed a psychoanalytic concept called the *after expulsion* of dreams that offered a possible explana-tion for this disappearance of the content of dreams from our mem-ories. We will have more to say about Freud's theories later, but perhaps the contentless dream phenomenon is only some under-standable part of normal memory and forgetting. I really don't know, but it is very interesting that in our studies, contentless dreams were more likely to be reported when one was not in one of these dysphoric moods. A relatively good mood when one goes to sleep tends to be followed by a feeling one has dreamed, but cannot recall it, or one has no recall of dreaming at all, while a bad mood tends to be fol-lowed by a dream one remembers well in the morning. Interesting! You might want to see if your recall for dreams follows this pattern.

Staying with dream recall, what about the usual suspects, age and gender? Do they make a difference?

Memory—particularly short-term memory—is usually not as good for older people as it is for younger people, so it is not surprising that dream recall is higher for younger people. Generally, women have better recall for dreams than men do. This gender difference is made clear by adolescence. One explanation for this gender difference may be the different experiences that girls and boys have in growing up. These differences may contribute to increased self-reflection.

Does motivation—really wanting to remember your dreams— make a difference?

Yes. If you are really interested in remembering your dreams, you are more likely to remember them. Many years ago my research partner Roland Tanck and I carried out a study in which we asked college students to indicate the level of their interest in dreams. We later found that this measure related positively to the number of dreams they subsequently reported.

How about personality differences?

Psychologists have developed a measure dealing with openness to experience. It was reported that people who score high on this measure tend to have higher levels of dream recall. The attempts to relate scales from more traditional personality tests to dream recall have not been particularly productive. For example, Roland Tanck and I tried to replicate findings from studies reporting some relationships between scales from the well-known MMPI personality test and dream recall and we found nothing. This shows the importance of replication in research.

Wouldn't the content of a dream make a difference in the likelihood of it being recalled? For example, wouldn't a really bizarre dream be more likely to be remembered than a mundane one?

In a word, the answer is *yes*. This was demonstrated in a sleep laboratory study.

If you want to remember your dreams, what can you do? Are there some easy steps to follow?

Perhaps the easiest thing to do would be to keep a notepad and pen by your bedside or perhaps even a small recording device so that when you awake from a dream you can record dream memories immediately. As most of us have experienced, dreams can be very ephemeral, disappearing from memory rapidly. Immediate reporting of the dream can be important, as distractions can easily wipe out the content of dreams.

A device called the *Nightcap* was developed to help collect home dream reports. This is a computer-based system that detects the presence of eye movements during sleep and then awakens the dreamer.

All of these techniques may be helpful for people who want to look more deeply into their dreams. I am not so sure that it would go over very well with your partner when you are sharing a bed, not if he or she is in need of a good night's sleep.

Given all that you have said, are there some people who do not remember dreams?

Yes. People who have stated that they do not dream have been studied in sleep laboratories. When some of these non-dreamers have been awakened during REM sleep, they will report dreams, but this is not true for all. One suspects that either their memory for dreams is so transient that recall is wiped out or they do not experience dreams even when conditions are ideal.

Do animals dream?

If you are a pet owner, and you watch your pet carefully while it is sleeping, you may have a more intuitive answer to that question than I do. Certainly mammals, great and small, experience REM and NREM periods of sleep, and it is a reasonable assumption that they probably experience dreams during REM sleep periods. Researchers are developing very small instruments to measure eye movement activity in small mammals to carry out research on sleep, so we should learn more about the biology that underlies dreams in these animals, but what do they dream about?

Creative artists have sought to fill this vacuum with wonderful whimsical ideas. If I remember correctly, the late beloved cartoonist Charles Schultz had everyone's favorite dog, Snoopy, dreaming quite a bit in his comic strip *Peanuts*. How much of these dreams were Snoopy's or Schultz's, we'll never know.

While the discovery that REM sleep was linked to dreaming and the fact that animals experience REM sleep provided evidence that animals very likely dream, the idea that animals dream preceded these discoveries. Walt Disney, who delighted audiences in the 1920s and 1930s with his cartoon characters Mickey Mouse and Donald Duck who acted very much like humans, also provided us with a dog who experienced dreams—and this was before the publication of the discovery of REM sleep by Aserinsky and Kleitman. The Disney movie with its dreaming dog was a version of the fairy tale *Cinderella*. The dream scene occurred in a barn where a goofy dog, Bruno, was sleeping and dreaming about his enemy, the cat Lucifer. Cinderella sees that Bruno is clearly asleep and seems to be experiencing a terrible dream in which he is chasing the cat. After some difficulty, Cinderella awakens Bruno and reprimands him for his aggressive dream.

While we really know precious little about the dreams of animals, research indicates that animal rights activists report animal characters in their dreams at a much higher rate than the general population, and the dream animals are overwhelmingly friendly.

How often do we dream in color?

Probably more than we think. At one time, it was believed that dreaming in color was rare. In fact, back in the 1940s, questionnaires were given to college students asking whether they dreamed in black and white or color, and nearly three-fourths of the students replied that they rarely or never dreamed in color. However, when a study was carried out in a sleep laboratory in the 1960s, the responses of the subjects were dramatically different. In more than three-fourths of the dreams reported, color was present. People may be less likely to report dreaming in color when they dream at home, because recall for dreams in the laboratory is almost immediate. At home, upon waking in the morning, it is often hard to recall the details of dreams and memory for color could be an early casualty.

Based on self-reports, younger people are more likely to report that they have had this experience than older people.

What are the most common emotions reported in dreams?

There is a line from a beautiful song which goes "Sweet dreams be yours, dear, if dreams there be." Unhappily, unpleasant emotions are reported far more often in dreams than pleasant emotions. The most common emotion reported is *apprehension*. As we will see later, the word *apprehension* covers a lot of ground, from concern that something will not turn out well to outright fear. Interestingly, anger, a negative emotion that we all feel from time to time, ranks relatively low among the feelings we experience in our dreams, but it does occur from time to time.

Can you give us an example?

Sure. Here is a fragment of a dream provided by a friend:

> I am sitting in an airline terminal and "God Bless America" is being played over the speakers. I am singing along heartily to it, and even harmonizing with the song…. Anyway, while I was singing, I noticed a guy in the row in front of me, and he's chatting with his wife while the song is being played. I was annoyed by the fact that he was not paying attention to the patriotic song, so on the final "home" of "my home sweet home," I sang it right into his ear. Well, as you can imagine, that did not sit well with him and he glared at me.

You can see the anger very clearly in the dreamer and another character, but apprehension is much more common in dreams than anger.

4

The Sources of Dreams
and the Experience
of Lucid Dreaming

What are the sources of dreams?

As we have suggested, the primary sources of dreams are your experiences. What may have happened to you yesterday and what may have happened to you years ago, and anything in between, could work its way into a dream. Freud, who was an acute observer, noted that the trigger for dreams was typically some event that had occurred in the past 24 hours or so. This event could be intertwined by a hypothetical mechanism Freud called the "dreamwork" with more distant memories to construct the dreams we experience. What Freud conceived as the "dreamwork," and what we would now try to specify as the action of certain areas of the brain, transforms these memories into a kind of theatrical production: a production that can be very different from the memories that gave rise to the dream. The types of experiences that stimulate a dream could range from watching a movie (has anyone ever told you that she watched a horror film and woke up from a nightmare that night?) to events you witnessed, conversations you had, and experiences that were important enough to register in your thoughts and memory.

In a classic experiment, researchers presented a stimulus to a man while he was experiencing REM sleep. One of the stimuli was cold water being sprayed on the subject's legs and feet. The researchers awakened the subject and found that in his dream report children came into his room and asked for water. He handed a glass of ice water to the children but in doing so he spilled the water on himself.

This very interesting experiment suggests that dream content can be influenced by actions carried out on a person while he or she is sleeping, but one can see immediately that this would be an impractical way to alter dreams in real life. One wonders whether one could do something to people before they go to sleep which might affect the dreams that may follow. Wouldn't it be nice instead of simply saying, "Sweet dreams" to a person before he or she goes to sleep to be able to do something that would increase the chances that the dreams actually would be sweet rather than unsettling? There are studies that suggest that priming people to have more pleasant dreams is possible, and as we shall see later, the idea has been successfully extended to alter the dreams of people who suffer from chronic nightmares.

Let's talk a bit about priming dreams. Experiments have shown that both the mood experienced during a dream and the content of the dream itself are susceptible to alteration by priming a person before he or she goes to sleep. Here is an example dealing with mood. In the study, students were first asked to keep a dream diary for a week to provide a control sample of dreams. During the next week, they were asked to look at a picture before going to bed that generated either positive, neutral or negative emotions. The following morning, the students recorded their dreams. During the nights in which they looked at the positive emotion-generating picture, their morning dream reports were more positive. During the nights they looked at the negative emotion-generating picture, their morning dream reports were more negative.

Now, if this simple priming device, a picture, was able to generate positive effects in dream content, we might wonder what the effect of more substantial priming on dreams would be; perhaps something not too exciting, like some kind, reassuring words before bedtime or even a soothing back massage. Try it sometime. If it leads to sweeter dreams, let me know.

Let's leave controlled experiments for a moment and return to Freud's observations that events taking place shortly before dreams seem to find their way into our dreams. Freud's belief that incidents during the 24 hours preceding sleep tend to become part of our dreams has been confirmed by studies in the sleep laboratory.

Incorporation of events into nighttime REM dreams tends to be highest on the day the events took place. Researchers have observed another very interesting pattern. There seems to be a second peak of incorporation of daily events into nightly dreams and this occurs about five to seven nights after the event happened. This has been termed the *dream-lag effect*. It is not clear why this happens, but it is instructive that researchers have reported that the dream-lag effect seems most likely to occur when the events are personally significant rather than simply reports of what one does during the day, i.e., the more routine actions that are part of our daily lives. Importantly, the dream-lag effect was not evident for dreams that occur during non–REM sleep.

When we delve more deeply into the memories that are the sources of dreams, research suggests that what we call *autobiographical memories* are the primary source for dream content. Autobiographical memories encompass your experiences both in the short term (what might have happened to you yesterday or during the week) and longer-term events that could date as far back as early childhood. Your autobiographical memories are something like a diary of your life, a diary that your mind may update and modify at any time. These memories are changeable views of your life, experiences that your mind constructs and edits, somewhat analogous to writing and editing a novel. Autobiographical memories may draw on the specific events that happened to you which, by the way, are called *episodic memories* (that game of softball you played last week, the discussion you had with a coworker a month ago), as well as the images and portrayals of events you absorb from watching television and movies, things that you were not personally involved in. Research has indicated that you are not likely to experience a video-like portrayal of specific episodic memories in your dreams. It is the more filtered, elaborated and mind-constructed autobiographical memories that are typically the components of dreams. In the dreaming process, your mind integrates and transforms these autobiographical memories into a new, different narrative, sometimes in a coherent fashion, sometimes in a more haphazard way.

As you dream, the actions that take place may be based on memories from your past experience that seem fragmented, unrelated and disjointed. Some dreams may appear quite puzzling. As we said, you are unlikely to re-experience an event that occurred to you in a given time and place just the way it happened. However, if you are asked to associate to the different events in your dream, a number of real-life incidents in your past will often come to mind, and I think you will see—as demonstrated in the second part of the book—that examining the meaning of these real-life experiences can often be revealing in thinking about your own personal psychology.

If autobiographical memory is the major source of dreams, and personally significant events can affect your dreams even a week after they happen, then would something that is happening to you that is really important find its way into your dreams? Say, something like getting married?

I have never seen any studies of dreams relating to upcoming weddings, but I think the point is well taken. If something is really monopolizing your thinking, I would imagine that it might have an advantage in finding its way into your dreams. Here are a few examples from the research literature which support this conclusion.

Women who were going through a divorce and feeling depressed were studied for about five months. The more concerns a subject expressed about her former spouse, the more often she dreamed about him.

A study carried out on women who were pregnant found that during the pregnancy, about two-thirds of the women dreamed about the pregnancy or baby. What I found very interesting was that about 40 percent of these women reported that at least one of these dreams was frightening, often upsetting. Interestingly, the most common dream reported involved conflict with the father of the baby.

The most recent research I have seen on the dreams of pregnant women confirmed the earlier findings that their dreams were not all that pleasant. In late pregnancy, dream content tended to be more morbid. In general, the emotional tone of dreams of pregnant women was characterized as more dysphoric than a sample of women who were not pregnant. As one might expect, childbirth content tended to be higher in the late stages of pregnancy than in the early and middle stages.

A third example of important events finding their way into dreams is developing cancer, which for many people can become an overwhelming problem and, as such, may become a central focus in their thinking. Studies of the dreams of cancer patients suggest that their dreams often deal with the patient's attempts to cope with the loss of self-sufficiency, increased dependency, disfigurement, and death.

It has been my view that dream content often reflects lingering, difficult problems that remain unresolved in the person's life.

I'd like to return for a moment to events that stimulate dreams. If you go to bed hungry or thirsty, would that influence your dreams? Would you dream about food if you're hungry or having something to drink if you are thirsty?

It's not too likely, but you might. Some people, not a great number, were studied in sleep laboratories after being deprived of water for 24 hours. About one-third of their dreams contained some imagery relating to thirst which is a lot higher than would usually be the case. So, being deprived of water appeared to have some effect on dream content.

The study of dreams of people with eating disorders bears some relevance to this question. A study carried out in a sleep laboratory indicated that people with eating disorders such as anorexia (eating very little food in an attempt to remain thin) or bulimia (eating and purging) dream about food far more frequently than people who do not have eating disorders.

Can eating certain foods influence what we dream about?

I have never thought much about this, but according to a very interesting paper by Tori Nielsen and Russell A. Powell, published in *Frontiers in Psychology*, a lot of people are believers. The authors noted that anecdotes about recurrent dreams, bizarre dreams, and nightmares triggered by eating a particular food or eating too much or too late at night are common. One of the foods that some people believe induce bad dreams is cheese. This idea has received so much attention that the British Cheese Board carried out a study with 200 volunteers to investigate the effects eating cheese has on dreams. According to Nielsen and Powell, the study reportedly did not find evidence that eating cheese prior to sleep caused nightmares, but did find evidence that different types of cheese could induce different types of dreams. For example, Stilton was associated with crazy or vivid dreams while cheddar was associated with dreams of celebrities. Nielsen and Powell noted that the study was never reported in a peer refereed journal, so it is difficult to evaluate its scientific merit.

If you are ready to give up cheese sandwiches and switch to, say, cucumber sandwiches, you should know that that an eminent psychoanalyst of some years back, Earnest Jones, noted the then-current belief that eating cucumbers could lead to nightmares. The most recent data that I have seen relating reports of eating certain foods to subsequent dreams was reported by Nielsen and Powell, who wrote that very interesting article just described. They gave out questionnaires to 396 students inquiring about the goods they ate and their dreams. The students were asked whether they had noticed if foods produce bizarre or disturbing dreams and if eating late at night had influenced their dreams. About one out of five of the students said that they did notice a relationship between food and dreams. The food most often cited as the cause of bad dreams was dairy products. Is this another vote for the cheese hypothesis? Obviously, these findings report students' beliefs, and observations are nothing close to a controlled experiment, so they can't tell us whether there is a causal relationship between eating dairy products and disturbed dreams.

Interestingly, student reports of disturbing dreams were correlated with reports of binge eating and eating for emotional reasons. Equally interesting, people who reported a healthier diet also reported more vivid dreams. So, at the risk of making a particularly poor pun, we could say there is indeed food for thought here.

The idea that food could cause bad dreams has found its way into literature. In Charles Dickens' memorable tale *A Christmas Carol*, Ebenezer Scrooge blamed one of his ghostly experiences on undigested beef, mustard, cheese, and a piece of underdone potato.

Continuing with experiences that might stimulate a dream, dreams are clearly visual. Suppose you were born blind and couldn't see. Would you still have visual content in your dreams?

The research I will cite is based on a small number of cases. Still, the answer seems to be an emphatic *yes*. People who were blind from birth reported visual content in their dreams. In many ways, this seems counterintuitive but it does appear to be the case. Not only did they report visual images, they were able to make drawings of their dream images. I am not sure exactly what is going on here, but this may be a testament to the power of the brain and our imagination to construct a world that cannot be seen. I wonder if this is similar in some ways to when we read a novel and are able to picture the setting and the characters in our minds. After all, this can be a place, even another planet, we had never seen and is entirely fictitious.

Not surprisingly, the dream reports of people born blind often include the input of other senses such as the senses of audition (hearing) and olfaction (smell). Researchers have reported that the dreams of the blind often include a lot of conversation. This should not be surprising, as conversation is a critically important tool for blind people to better understand and cope with the world about them. Researchers also reported that the dreams of the blind often took place in the familiar environs of their homes.

You have said that pre-sleep priming may have some effect on dream content. Is it possible for dreamers themselves to significantly affect dream content as they sleep?

Maybe. Researchers have been studying a phenomenon called *lucid dreaming*. When I first heard the term, I thought it was a description of dreams that were extremely vivid or colorful or seemed exceedingly real. While some of these things might be true, the essential meaning of the term is that the dreamer is aware that he or she is dreaming and sometimes can influence what is happening in the dream. Surveys suggest that most people believe they have had this experience at one time or another. In a representative sample of German adults, just over half (51 percent) reported that they had experienced a lucid dream at least once. A recent review of lucid dreaming studies carried out over the years produced a very similar figure, so it looks like about half of us have had this experience at one time or another. Think back. Have you ever had the experience of modifying what was happening in a dream, stopping something or changing things around? If so, have you had this experience frequently? Say, once a month or more? The figures offered from research are that about 23 percent of the people asked responded yes.

So while lucid dreaming may be common, I would be a bit skeptical of the figures reported as they are largely based on questionnaire data. Interestingly, there is a strong relationship between how often people report they recall their dreams and their report of experiencing lucid dreaming.

Researchers have suggested that lucid dreaming is a hybrid state of consciousness with similarities to and differences from both waking consciousness and REM sleep. We shall look at a recent study that explores this hybrid state, but first let's take stock of two of the difficulties in conducting such research. The first problem is finding subjects who can really demonstrate that they have this capacity. The second is communicating with a person who is in the midst of lucid dreaming. The person is sleeping, and if you wake him or her, the phenomenon you are studying vanishes.

While statistics indicate that there are lots of people who report that they are lucid dreamers, when you test them in a sleep laboratory, they are unable to demonstrate this ability. In one study, for example, researchers recruited 20 subjects who said they were lucid dreamers, but in the laboratory only a handful were able to provide clear-cut evidence of lucid dreaming.

The problem of communicating with the person while he or she is dreaming has been resolved, at least in part, by the ability of lucid dreamers to make prearranged eye-movements that can be picked up electronically. These are not precise signals translatable like Morse code, but it is discernible communication, albeit on a primitive level.

Here is a brief description of research carried out by a German team trying to learn more about the activities of the brain during lucid dreaming:

To study lucid dreaming, the research team recruited four men through the Internet who said they were experienced lucid dreamers and had been so for many years. The men slept in an MRI machine so areas of brain activation could be studied while their eye movements and other physiological data were monitored by other devices. To trace the onset and fading of lucid dreams, the men were given the instructions that as soon as they achieved lucidity in their dreams, they should move their eyes to the left, then to the right, then left again, then right again (LRLR), and then after this to clench their left hand for about ten seconds. They were told to repeat these LRLR eye movement signals, and this time clench the right hand, and to continue doing this sequence as long as possible. When these signals ceased, the researcher would wake the participant to inquire about his testing experience and obtain a dream report.

This all sounds very experimental, with lots of sophisticated equipment being constantly monitored, but put the details of the experiment aside for a moment and take a flight with your imagination. Try to picture a person just like yourself sleeping and suddenly beginning to experience an REM sleep dream. Then, as pre-instructed, the person consciously moves his or her eyes back and forth, letting the researchers know that he or she is dreaming and apparently able to alter the contents of his or her dream. Doesn't it sound a bit like a science fiction movie? The important thing, however, is that one of the four subjects recruited for the experiment was able to do at least part of this, to dream and let the experimenters know that he was indeed dreaming. There was a scene change in one of his dreams, but the research report does not tell us whether it was instituted by the subject or occurred without his efforts. That would be very interesting to know.

Lucid dreaming under the pressure of having to perform in a sleep laboratory is not easy to do. Only one of these very experienced lucid dreamers was able to carry out the instructions successfully, so we have a sample of one, which, of course, is as small as we can get and we should exercise appropriate caution about generalizing results. Still, the results of the experiment are very interesting, and let us briefly summarize what they found.

Periods of lucid dreaming which occurred during REM sleep were compared with periods of REM sleep in which lucid dreaming did not occur. Here are two observations from the research report: the first is that lucid dreaming was associated with a reactivation of areas normally deactivated during REM sleep. These findings offer an explanation for the recovery of reflective cognitive capabilities that we see in lucid dreaming. *Reactivation*, I think, is the critical word. The second is that the strongest increase in activation during lucid dreaming occurred in the precuneus, a region of the brain involved in self-referential processing. Stated more simply, when a person is experiencing lucid dreaming, there is some restoration of one's usually suppressed ability during dreaming to reflect on one's own mental state.

Does lucid dreaming have any practical value? An important implication from lucid dreaming is that it may be possible to alter one's dreams to make them more pleasant. This could be particularly beneficial if one experiences unpleasant dreams or nightmares. The question that then arises is can one be trained to become a lucid dreamer?

A variety of methods have been tried to help people become lucid dreamers. They include both training before sleep and efforts by the dreamer to develop these skills during sleep, going through a kind of practice session. A review article looking at these procedures suggests that the training methods now being used to induce lucid dreaming are not all that reliable. The reviewers looked through 35 published studies, evaluating the evidence for the effectiveness of different induction techniques. They concluded that while some of the techniques appeared promising, none of the induction techniques had been verified to consistently and reliably induce lucid dreams.

While the science isn't there yet to confidently inform us where training for lucid dreaming may be of real value, and how best to carry out lucid dreaming training, if one surfs the Internet, one will find that there are already institutes and academies that offer training for developing skills of lucid dreaming. Personal accounts and educational materials are available as well. Advice to the reader interested in lucid dreaming? Read the review study before opening your wallet.

5

The Content of Dreams, Theories About Dreams and Dreams That Recur

Lucid dreaming sounds interesting, but let's return to the dreams of the average man or woman who rarely experiences lucid dreaming. What are some of the typical dreams they experience?

While there are probably some types of dreams that are experienced universally, the culture that we live in probably influences dream content. A good example of the influence of the culture one lives in on dream content can be found in a dream book written on papyrus that survived from ancient Egypt. The book contained more than 100 dreams. The dreams dealt with such activities as carving, brewing, weaving and plastering, which reflect the importance of manual labor in the ancient world. Something very specific to ancient Egypt was a note indicating that a dream of eating crocodile flesh was thought to indicate that the dreamer would become a village official.

In their studies of different cultures, anthropologists have collected dream reports from a number of less technologically developed societies. These reports tend to show that the daily experiences in these different cultures are reflected to some extent in the dreams that people report. For example, dreams collected from American Indian tribes, in which hunting was important in obtaining food, contained more mentions of animals than dream reports collected from American Indian tribes that relied primarily on agriculture for their food.

When we turn our attention to the dream reports of people living

48

in modern, technologically-based societies like the United States and consider typical dreams, the question that poses itself immediately is who to ask. People recruited to spend a few nights in sleep laboratories? Or samples taken from the population at large? While the dreams we obtain from sleep lab reports are probably more reliable than those that we would obtain from home dream reports, the sample of subjects is very small and is unlikely to be representative of the nation as a whole. Therefore, I would be hesitant to make generalizations from such studies.

One alternative might be to use larger samples such as one might find in college or university classrooms, or to recruit people via the Internet, asking them to record home dream reports for a period of time.

Another alternative—one that I have the least confidence in— is to interview people or ask them in questionnaires about the dreams they have had over the years. The problem with this approach is that memory for dreams is notoriously poor. Asking whether you had a certain type of dream months or years after you may have had the dream is likely to yield less accurate results than asking you to keep a morning-after dream log. Recognizing this limitation, researchers have carried out such studies. Imagine a questionnaire asking, "Have you ever dreamed?" then listing a number of dream types with a "yes or no" response. Included on the list would be dreams we might consider typical—flying, finding money, being naked or partially dressed and feeling embarrassed, losing teeth, becoming lost, taking an examination at school, failing an examination, being chased, falling, and being unable to find a restroom. Researchers have given lists of "typical dreams" to hundreds of people. They have found that the odds that people will report such dreams at least once during their lifetime is very high. Moreover, if you show the same list to a different group of people several years later, you are very likely to find the same results. The typical dreams that people reported a few years ago are pretty much the same that people will have today. The typicality of dreams seems to be a stable phenomenon.

This does not mean, however, that people will have typical dreams very often. The results from data we have on this question seem mixed. In some studies when people were asked to keep a sleep log and report their dreams for a week or two, the frequency with which they reported such dreams was not all that high, and when people were studied in sleep laboratories, reports of "typical dreams" were almost non-existent. So while most of us may have such dreams at one time or another, these studies suggest that dreams of flying, falling, etc., really did not occur all that often. Case in point: German researchers investigated the prevalence of flying dreams in nearly 6,000 adults. Less than eight percent of the participants reported having such dreams within the last few months and this represented only a slight increase from past reports.

That said, recent research conducted on Chinese students indicated that most reported a "typical dream" among their recent dreams. Perhaps these different results reflect the way typical dreams were defined in the studies. The longer the list of "typical dreams" presented to the subjects, and the more inclusive the definitions are, the more likely a typical dream will be reported.

Even though dreams of flying do not occur very often, they still fascinate me. What does a dream of flying mean?

If dreams of flying through the air fascinate you, you are not alone. People have speculated about this for years. A dream book, perhaps 3,000 years old, found in the ruins of the ancient land of Assyria, offered interpretations for dreams of flying. Judging from what was inscribed on these ancient tablets, some of the would-be flyers apparently dreamed that they had grown wings. According to the Assyrian dream book, the consequences of experiencing dreams of flying depended in part on one's status in society. An important person would experience different kinds of events than a commoner. That's safe enough to say with or without a dream. Here is a prediction about flying dreams from the Assyrian dream book that may bring a smile. If you happened to be in jail (I really didn't know ancient Assyria had jails, but apparently it did), and if you had a dream of flying, it meant you would be freed. Does that bring to mind the get-out-of-jail free card in Monopoly?

I have seen only a few contemporary reports of dreams of flying. I can't provide any data on how often dreams of flying involve unaided flight or perhaps the use of homegrown wings in the Assyrian manner, but here are examples of some contemporary twists on the age-old idea of flying dreams. A retired scientist told me that when he was a preadolescent he had dreams of riding his childhood tricycle and it would lift into the air, taking him as high as the treetops. He said that the ride through the air was very pleasant. Another person, also retired, told me about a recent dream in which he discovered a button attached to his arm, perhaps sewn in surgically. When he pressed the button he experienced a short vertical lift into the air. The ascent was slow and pleasant and he hovered there for a while, pressed the button again and returned to the ground.

Thousands of years after the baked clay tablets of the Assyrian dream book had crumbled into shards, we still find people offering unproven ideas about the real meaning of dreams of flying to whoever will listen. Some of these ideas seem pretty fanciful to me, but to be honest, no one really knows whether there are any general meanings for dreams of flying. I have never seen a scientific study carried out on the question and I don't expect to see one anytime soon. In the absence of any serious research, people have been free to speculate and all sorts of ideas have been proposed.

The psychoanalytic community has offered its interpretations. Not unexpectedly, some researchers have seen a connection between dreams of flying and sexuality. Some thinkers in the psychoanalytic tradition have tried to apply psychoanalytic theory to characters in literature (e.g., Hamlet), and I wouldn't be surprised to learn that psychoanalytic writers might have drawn on the Greek myth of Icarus in their interpretations of dreams of flying. Remember the story? Icarus tried to escape from the island of Crete using wings made from feathers and wax but flew too close to the sun, causing the wax to melt, and he plunged to his death into the sea. The noted psychoanalytic writer William Stekel believed that there were death symbols in some dreams of flying.

There are countless sites on the Internet eager to tell you what dreams of flying really mean. My own biases in this matter are perfectly clear. I suspect that some of the stuff you might come across in an Internet search will be the rough equivalent of junk mail, unsubstantiated opinion masquerading as knowledge.

Dreams about taking tests and exams are one of the typical dreams that people have. Sometimes people have dreams about taking examinations years after they have completed school. Is that the case, and, if so what does it mean?

Once in a blue moon I have dreamed of taking examinations, and it has been decades since I have taken an examination of any kind. Often, these dreams have you running late for class, trying to find the classroom but unable to, missing the exam entirely, or arriving but not performing well. Recently I had a conversation with a retired professor who told me she often had such dreams and I wondered whether this might be an occupational hazard for people whose career training included many years of college and postgraduate work. The fact that dream content is largely based on autobiographical memories offers a rationale for this testable hypothesis.

Incidentally, the professor's dream was interesting because it added an element of the flying dreams we just considered to her prototypic examination dreams. Briefly, in her dream, she was back at the university she attended as a graduate student and was late for an exam. She wasn't sure which building she was to go to, and in her anxious search, she walked into one building and hurriedly checked out the classrooms on the first three floors. She drew a blank; nothing there. While on the third floor, she walked over to a window, looked outside and spied what she thought was the right building. Too far away to get there on time, she thought. The only way to get there was to fly.

As she related the dream, she wasn't sure what happened next, but she believes that as she was about to take the leap, she woke up.

What do retrospective dreams of examinations mean? One could offer more plausible explanations for what they might mean than for dreams of flying, which seem far-fetched at best, but objective research about examination dreams, as far as I know, is totally lacking. When we move into the second part of this book dealing with the analysis of dreams, and especially when we discuss the Dream Incident Technique, I will offer some ideas that may be useful in probing into what such dreams may be suggesting to us in our individual lives.

We talked some about the possible effects of eating and drinking on dreams. How likely is it that people actually do dream about food?

Not too often. The percentages reported depend on the way the data were obtained. Figures cited are as low as 1 percent when researchers looked at samples of actual dream reports. When people were asked whether they ever had a dream about eating delicious food, the figures go up significantly. Dreaming about a great meal may be a memory that sticks with you, particularly when the company and setting are special. A good illustration of this is a dream told to me by a person whose friends and relatives had moved far and wide, even out of the country. She dreamed she was enjoying a wonderful outdoor barbeque that included many of the people who had moved away.

Here is an example of a food dream I think you may find interesting because it foreshadows a topic we will consider later when we discuss really bizarre dreams. The dream which begins in a restaurant with the innocuous act of ordering a cup of coffee, is immediately replete with incongruent activities, and then features a rapid scene change, both of which are features of bizarre dreams. While the dream contains improbabilities, there is nothing dire or sinister in it. If anything, the tone resembles the slapstick comedies that earlier generations were treated to in silent films, most notably in the actions of the comic genius Charlie Chaplin:

> I was at a coffee shop waiting and waiting for my cup of Joe and there was a woman on a skateboard-type apparatus, more like a flat platform on wheels. There were kids on them, too, zipping all over the store. I was concerned that there might be an accident—especially mixed with hot coffee—so I said to the woman, "Are you sure you should be using this in the coffee shop?" She gave me a withering look and then zipped past me.
>
> So after a long wait still no coffee. The cashier was sitting behind the counter and I didn't realize she was on the phone so I went up to her and started yelling at her, something like, "We've been waiting a long time for coffee. Where is it?" Then I realized she was on the phone and I said, "I hope you are calling someone about the coffee. Is your machine broken or something? Why didn't anyone tell us? Where is your manager?" I don't remember her response because the next thing I knew, I was at a car dealership getting coffee from its free display. And they also had free

(and this sounds so good—I never thought of it in my waking hours) pecan rolls with CHOCOLATE GLAZE on them!! Yummy! Then I woke up because my phone was ringing.

The rolls sound good, don't they? Let's talk about love. Over the years, many popular songs have variations on the theme of dreaming about you. Is it true that people dream of their romantic partners?

They certainly do. For some people, these dreams can be quite positive, and for others, much less so. The content of the dream tends to reflect the kind of attachment the couple has in their waking life. Couples with feelings of security about their attachment tend to report dreams that reflect this sense of security.

How much explicit sexuality do we find in dream reports?

In a study carried out some years ago, researchers examined the dream reports of a fairly large number of men and women and found that about 12 percent contained sexual thoughts and/or actions.

Here is comparative data that may add perspective to these findings. An anthropologist reported that when he asked the men living in a tribal society in the Amazon about their dreams, the figures for explicit sexual thoughts and actions were not very different from what the American researchers found in their study. The figures for the Amazonian women were a bit lower than for the American women. In assessing this comparison, I must note that the researchers reported that both the men and women in this remote tribe were not particularly inhibited about sex. Both men and women were said to have had multiple lovers. As an incidental observation, two societies with radically different cultures, both with easy access to sexual stimuli, had similar frequencies of sexual-related dreams.

Let me throw in one more study that is more recent. Researchers collected a very large number of home dream reports from Canadian men and women. They found that about eight percent of everyday dream reports included some form of sexual activity. The most frequent type of sexual activity in the dreams was sexual intercourse, with somewhat lower figures for kissing and talk about sex. The women in the study were able to identify current or former partners in about 20 percent of their sexual dreams. The figure for men was slightly lower: 14 percent.

So, I think it's fair to conclude that about one out of ten dreams of the people who have been studied contain sexual content, and there is little indication that currently there are pronounced differences between men and women in this regard.

I have heard that Freud thought sex impulses underlie most dreams, but these impulses would have been suppressed. Would Freud have been surprised by the frequency with which sexual content appears in our dreams today?

I think so. Take a look on the Internet at any reference that shows photographs of women in the late 19th and early 20th centuries either in Europe or in America. Look at the clothes women wore. Compare them with the clothes women wear today. Or, better yet, take a stroll on one of our beaches during the summer. Compare the treatment of sex in novels written in Victorian times with novels written today. Compared to our own time, Freud lived in an era of suppressed sexuality and it would have been considered very unladylike to report such dreams.

Still, Freud believed that sexuality was a driving force behind most dreams even when there was no mention whatsoever in the dream report of sexual activity. He came to this conclusion by asking people to tell him their free associations when thinking about the various parts of the dream. He believed that the actual dream report—which he called the *manifest content*—was a kind of outlet for impulses that were hidden from the dreamer, impulses that were often sexual in nature. Even so, if these impulses were allowed into the dream content, they had to be disguised. Freud conceived of a kind of agency of the mind that he called the *dreamwork* to carry out this disguise. Working unconsciously, the dreamwork used several mechanisms to accomplish this objective, the use of symbols being one of them. He believed that hidden sexual impulses could show themselves as ordinary objects and actions in the dream, which in his view could really be symbols of sexuality.

Is there substantial scientific evidence to validate Freud's theories?

Not really, not in the sense of having carefully carried out controlled studies that can be replicated by many experimenters. People who adhere to Freud's theories in the psychoanalytic movement sometimes argue that they see evidence for the validity of his theories in the successful treatment of their patients, but that argument seems unconvincing to me, as it does not offer a real control for the nonspecific benefits of almost any type of psychotherapy. People can get better by simply talking with any sympathetic therapist, Freudian or not.

Speaking of symbolism in dreams, is there any evidence that specific objects or actions in dreams have any meaning beyond what they appear to be in the dream?

The idea that objects or actions in dreams may be conveying hidden meanings goes back a long way. Scholars studying the civilizations of antiquity—in Egypt, the Near East and Greece—can point to examples of dream interpretations in which the objects in the dream were thought to represent something else than what they appeared to be in the dream. For example, in ancient Egypt, a shining moon was said to represent forgiveness while a large cat was said to represent a bumper crop. The latter interpretation recalls the biblical story of Joseph interpreting the pharaoh's dream. If you think about it, for many people in our own time, not all that much has changed since the days of the pharaoh's soothsayers and the divinations carried out at the Oracle of Delphi. They would readily accept the idea that specific objects or events in dreams have readily identifiable alternative meanings. I have dim memories of the time you could walk up to a newsstand and buy dream books which provided a this-means-that version of dream analysis and, I suspect, dabbled in numerology as well. You can buy books called dream dictionaries that offer alternative meanings for such commonplace objects as houses, shoes, umbrellas, and bodies of water in your dreams.

Is there any validity to these proclamations? Who knows? Robust scientific studies testing these ideas would be difficult to carry out, and if there were such studies, I suspect they would probably be ignored by many people who are intrigued by such things.

My attitude toward Freud's theory of dream symbolism is a bit different from the more wide-ranging this-means-that approach. Freud had a comprehensive theory about the pressures exerted by hidden, unacceptable impulses and the role dreams could play in affording an outlet for them. In his seminal work, *The Interpretation of Dreams*, Freud provided a list of symbols that could stand for male and female sexuality. It should be stated that Freud pointed out this was not always the case, that sometimes objects in dreams were exactly what they seemed to be and there was no need to go beyond that. Freud postulated that male sexual symbols in dreams would be clearly elongated objects such as tree trunks, umbrellas, sticks, and snakes while female symbols would be containers such as boxes, cases, chests, and hollow objects.

While neither my research partner Roland Tanck nor I were Freudians, we were intrigued enough by the theory to take a serious look at it. We hypothesized that people who were experiencing a satisfying sexual life would feel less pressure from ungratified sexual impulses and would show less evidence of these sexual symbols in their dreams than people whose sex life was essentially non-existent. We carried out a study to test this proposition. We asked university students about their dating life and compared two groups—those who dated frequently and reported a satisfying relationship with a person of the opposite sex and those who hardly ever dated and had no satisfying relationship. We collected dreams from the students, searching them for the presence of the symbols that Freud had listed as representing sexuality.

Here are two excerpts from the students' dreams that we classified as containing sexual symbols using Freud's list of symbols as a guide:

> Driving down the road. It was very snowy and icy.... A cement pole had ice and snow on it.
>
> I was back at my old high school. I was just visiting.... When I came out of class, for some reason I started looking for a locker. All of the lockers in the hall except one had a lock on it. I went to the open one and started to clean the things in it out. The locker number was my old girlfriend's locker. All of the things in the locker belonged to her. I took them all out and put my things in.

I think these examples would have brought a knowing smile to Freud.

The real question, of course, is whether the students who were not dating and did not have a satisfying relationship had more of these symbols in their dreams than the people who were dating frequently and had a satisfying relationship. The answer surprised me. *They did.*

We were still quite skeptical, which is a hallmark of good science, but our interest was sufficiently piqued to run another study. We decided to test a hypothesis advanced by a group of psychologists who stated that "symbolic expression of sexual arousal is required to a greater extent in persons whose anxiety is high, since such individuals are more likely to inhibit the direct or manifest expression of sexual arousal." Anxious people should use more symbolism. Interesting idea.

We began by looking at data left over from the study we just described. We had collected baseline data on the level of anxiety the students had been experiencing and related this to our measures of Freudian sexual symbols from the dreams. Again to our surprise, we found that *yes indeed* there was a positive relationship between anxiety level and symbolic sexual dream content. Once more in the spirit of scientific inquiry, we decided to replicate the study to see if our finding had real staying power. What we found in our second study was that the overall student anxiety levels assessed over a ten-day period were positively related to symbolism in the dream reports during that period. This replicated the results using the data from our original study. However, we also found there was no direct link between the appearance of symbolic dreams and the level of anxiety the students felt during the day preceding these dreams. It was the general level of tension the students were experiencing that seemed important, not how anxious they were that day.

We undertook one more analysis with data from our second study. We decided to look at the manifest sexual dream content and relate it to anxiety level. Here we included explicit references to sexual intercourse, petting, kissing, nudity, and erotic desires. We found a positive relationship between the level of anxiety the students were experiencing and instances of these explicit sexual actions in their dreams.

Positive results again, but not exactly what the hypothesis would have predicted because it assumed that sexual symbolism in dreams was a substitute for actual sexual expression. What we found instead was that both manifest sexual content and Freudian dream symbols were more prevalent in the dreams of people who were experiencing anxiety.

I must admit I don't quite know what to make of all this. To this day, I remain skeptical of Freud's theory of dream symbolism. I believe there may be other explanations for what we found, but the data is interesting and I invite readers who would like to try their hand at being a psychological theorist to see what they can make of our basic findings: *higher levels of anxiety and a more restricted sex-life both appear related to dream content that Freud theorized could serve as symbols for sexual experience.*

Two studies in psychology are hardly definitive proof, so in my papers describing the research, I certainly did not declare a victory for Freudian theory, but this little drop in the well of ignorance about the validity of symbols certainly made me wonder whether there may be something to what Freud was talking about.

I have heard things from time to time about Freud's theory and you have given us some of his ideas about dreams. Could you tell us more about Freud's theory?

Freud's book *The Interpretation of Dreams* has been the most important work written about dreams from the time of its publication in Germany in 1900 to the discovery of REM sleep in the laboratory in the 1950s. It is probably less influential today because we understand a great deal more about sleep and dreams than in Freud's day. Nonetheless, it remains influential, particularly to those in the psychoanalytic movement who use his theories as a basis for carrying out psychotherapy.

Freud was born in 1856 in the town of Freiburg, then part of the Austro-Hungarian Empire. His family moved to Vienna, and when Freud reached the age of 17, he became a student at the University of Vienna where he studied medicine and became a neurologist. He became interested in hysteria. In those days, hysteria was not thought of solely as someone going to pieces under stress or acting almost theatrically rather than calmly. Hysteria was the name given to an inexplicable disease in which the person afflicted (usually women) lost the ability to see or hear (hysterical blindness and deafness) or the ability to use their hands, arms or legs (hysterical paralysis). Such loss of normal functioning could not be explained by any disease, accident, or injury to the body. It began to look as if the problem might be psychological in nature.

The foremost authority on hysteria at the time was a French physician, Jean-Martin Charcot, who practiced in a hospital in Paris. Charcot gave lectures and demonstrations on how he successfully treated cases of conversion hysteria using hypnosis. One can see photographs today of Charcot talking to a group of men listening attentively while a patient suffering from conversion hysteria was being held and protected from falling by another physician. Freud traveled to Paris to observe these demonstrations and was very much impressed by Charcot. When Freud returned to Vienna, he tried using Charcot's methods himself but was dissatisfied that his cures for hysteria were transitory and incomplete. He began to try a different approach, carefully listening to his patients, occasionally voicing comments, and he developed what would became a talking cure and eventually the theory and practice of psychoanalysis.

A central tenet of psychoanalysis is that part of the mind is unconscious. It is a reservoir of ideas, thoughts, wishes and emotions we are normally unaware of. Dreams were important because Freud felt that the unraveling of dreams provided a "royal road" to the unconscious, a way of identifying what was lurking below the surface and of bringing these ideas to the patient's attention. When fully understood and accepted by the patient, these ideas could relieve symptoms such as hysterical paralysis. Over the years, Freud developed and refined his model of the mind and developed a following of both doctors and laypeople who treated patients experiencing a variety of problems.

It is hard to summarize Freud's elaborate and often technical theory of dreams in a few words, but in essence, Freud believed that wishes were the driving force of dreams. They could be for the attainment of a variety of desires, but in Freud's view, they were very often sexual in nature. Such wishes, often unacceptable in Freud's time, were relegated to the unconscious, but they were so strong they would try to gain expression in a variety of ways, particularly in dreams when an individual's conscious controls were relaxed. Even then, though, these desires were met by a countervailing force Freud called the *censor*. What emerged in the dream was a kind of compromise between these psychical pressures, with the unacceptable wish being disguised by the dreamwork into something else, so the true meaning of the dream would have to be disentangled and clarified, ostensibly by someone trained in Freud's psychoanalytic procedures.

Freud believed that the dreamwork could do a number of things to make the real meaning of the dream obscure. For one thing, it often combined the unconscious thoughts seeking expression with recent experiences, usually from the past 24 hours. It could also change the intensity of the original dream thoughts, turning a powerful idea into one that was trivial. Freud called this process *displacement*. The dreamwork also condensed a lot of suppressed ideas into a few fragments, a process he called *condensation*. Finally, as we saw earlier, the true meaning of the dream could be couched in *symbols*.

Freud had another idea about dreams that, while not central to his theory, is nonetheless interesting. He thought dreams kept people sleeping when they might otherwise have woken up. Dreaming allowed people to continue sleeping. He called dreams *the guardian of sleep*. The obvious problem with pushing this theory too far is the *nightmare*. It would take a very convoluted theory to rationalize how having a nightmare would help one stay asleep. The nightmare, by most definitions, includes a sudden awakening. A frightening dream helps you stay asleep? That would really be squaring a circle, and Freud, who liked to extend the reach of his theories to interpret many human endeavors, would not attempt to cross that bridge too far.

Are there any other theories about dreams besides Freud's?

Yes. There are theories advanced by other members of the psychoanalytic movement such as Carl Jung and by contemporary dream researchers, and many if not most people have their own ideas about the nature and meaning of dreams. Some years ago, Roland Tanck and I asked university students whether they had any beliefs or theories about the meaning of dreams. If so, what were their beliefs?

Eighty percent of the students answered yes. A number of ideas were mentioned—dreams are responses to pre-sleep stimuli; related to events of the preceding day or recent past; attempts to solve problems; or related to fears and anxieties. In shades of the thinking of the ancient civilizations of the Mediterranean, some students thought dreams could predict the future. When we asked another group of students which of these ideas were true or probably true, the statement that was endorsed most often (91 percent) was dreams are related to one's fears and anxieties.

Freud's theory of dreams probably influenced some of the students who thought that dreams were products of the unconscious and that dreams expressed wishes that were not gratified in reality. When we think of the general public, I think it is fair to say that Freud's theory of dreams has been, by far, the most influential theory of dreams and an integral part of his overall psychoanalytic theory.

Did Jung differ much from Freud in his thinking about dreams?

Yes. Carl Jung was Freud's protégé and then rival in the psychoanalytic movement. Jung had a basic disagreement with Freud about dreams. Dismissing Freud's idea that dreams were disguised wishes, Jung said they were nothing other than what they appeared to be. Jung postulated that one of the most important functions of dreaming was to bring up ideas that a person either was not aware of or had not been paying sufficient attention to in daily life. He called this tendency for dreams to bring up unattended matters *compensation.* The idea that dreams were compensatory was a major focus in Jung's approach to dream analysis.

Recall that Freud developed the concept of the unconscious and that it was integral to his theory of dreams. Jung accepted the idea of the unconscious but over time expressed major differences with Freud about the nature of the unconscious. In addition to an individual's unconscious (which, by the way, Jung called the *personal unconscious*), Jung developed a concept he called the *collective unconscious.* The collective unconscious was a structure in the mind that did not develop individually based on the uniqueness of a person's experience, but was inherited from our earliest human ancestors, an inheritance we all share. The collective unconscious consists of pre-existing forms Jung called the *archetypes* which we can only discover and study indirectly. These archetypes, sometimes described as motifs, elements, or mental images, were said to be universal, archaic and primitive. Some scholars have been fascinated by Jung's ideas about the collective unconscious and have searched for archetypal images in mythology and dream imagery. There are other people, myself included, who find the idea of a collective unconscious filled with archetypal symbolism extremely murky, and when applied to the study of dreams, not easily amenable to scientific research.

Neither Freud nor Jung, brilliant though they were, nor the earlier researchers who tried to carry out objective studies about dreams, had the benefit of the advances in understanding that came with the development of sleep laboratories in the 1960s.

What are some of the modern ideas about dreams, ideas that have come into currency after the development of sleep laboratories?

People are continuing to develop interesting ideas about dreams. Here are three examples:

When I discussed the findings related to the selective activation of areas of the brain during REM sleep, I briefly speculated about the possible evolutionary value that a combination of anxiety in the dream and ideas that emerged for problem solving might have for our ancestors who lived in a high-threat environment. Some contemporary theorists have developed a theory that highlights the possible value of anxiety arousal in dreams. It has been suggested that dreams have the biological function of simulating threatening events. The dream replays both the recognition of threats and of ways to avoid them. The theory proposes that our dream-production mechanism favors the selection of threatening waking events and replays them repeatedly in our dreams, and in doing so has survival value for the dreamer. In this view, dreams are something like a kind of drill, a dress rehearsal, preparing the dreamer to respond more effectively to the threats he or she might encounter during waking hours. The dreams of children, recurrent dreams, and nightmares have been offered as examples of this possible coping function of dreams.

A second theory of dreams is the "activation-synthesis hypothesis" developed by J.A. Hobson and his colleague. The research draws on the research of past decades following the discovery that dreaming is related to REM sleep and to subsequent studies investigating links between brain activation and dreams, issues we have discussed previously. The activation-synthesis theorists believe that dreaming is *physiologically* determined, a far cry from what psychoanalysts like Freud thought—that dreams were derived from psychological issues such as hidden wishes. Remember Freud's concept of the dreamwork, an agency of the mind that shapes the content of dreams? The dreamwork, with all of its hypothetical psychological mechanisms such as condensation and displacement? Nothing like that here. The activation-synthesis hypothesis argues that dream sleep is generated by a neural mechanism in the brain stem. This mechanism is believed to be preprogramed, something that is part of our biology, a kind of self-starting engine that fires up after sleep, not only switching on our dreams, but playing a significant role in shaping them. The concept of this neural generator of dreaming was based on observations of the periodicity and automatic features of forebrain activation. While a flurry of neurologic activity is hypothesized to be involved in this dream generating process (e.g., sporadic firings of reticular, vestibular, and oculomotor neurons), the analogy to a self-starting and self-stopping machine is hard to resist.

Given this surge of neural activity, how do dreams come about? The proponents of this theory believe that the activated forebrain compares information from specific brainstem circuits with stuff that is stored in our memories (presumably autobiographical memories) and synthesizes it all into a dream. They believe that the activated forebrain plays a significant role in shaping the dream, particularly the time and space contours of the dream and the rapid forgetting of the dream. The activation-synthesis hypothesis posits correspondences between the physiology of the brain and our subjective experiences of the dream.

While this is a complex theory, I believe the gist of the idea is that dreams are essentially a physiological phenomenon, the product of considerable neurological activity, largely fomented by the brainstem. At a minimum, the theory seems to downplay the role of psychological factors in shaping our dreams, and perhaps even shoving psychologists like me who believe that dreams have meaning for the individual—and poets and songwriters who use the romanticism in dreams in their creative work—into the dustbin of history.

Do I believe in this theory? The short answer is *no*. To my mind it conflicts too much with the findings of my own research, particularly the studies I will discuss later using the Dream Incident Technique, and with my clinical experience, both of which suggest that dream content is psychologically based, reflecting our experience and particularly unresolved aspects of this experience. I am very skeptical that such dreams and their relations to real experience could be a preprogrammed product of our brain stems.

A third view expressed by some contemporary theorists is that dreams are an extension of the activity of waking life. The theory holds that our waking states—our experiences and concerns—are reflected in our dreams. This approach, which you might rightfully surmise I like, has been called the *continuity hypothesis* of dreaming. People espousing this position have tested it by asking subjects to keep dream diaries for a period of time, perhaps two weeks, and simultaneously report on their waking activities. The researchers have found that subjects' reports of the amount of time they spend in waking life activities tends to be reflected in the content of their dreams.

In developing the theory, I think it is fair to ask what areas of waking life are likely to give rise to dream content. My own view is that when one probes into the real-life experiences (both in the near past and more distantly) that are associated with the dream, one is led to the hypothesis that this is not a random selection of experiences. In my view these experiences are often unresolved issues, events, ideas, and needs that may still be bugging the dreamer. I believe that if the dreamer can trace these experiences and then ponder their significance in his or her life, he or she may be able to better understand the unresolved problems that are still a source of concern to him or her. Approaching dreams this way may enable people to better grasp what is still troubling them. As we shall see later in my discussion of the Dream Incident Technique, it is possible to formalize these ideas into a method of dream analysis which lends itself to systematic and objective research.

If dreams have meaning for the individual, either looking at the content of the dream itself or at the experiences that are associated with it, wouldn't it make sense that if one has recurrent dreams, that the themes present here would have particular significance for the dreamer?

It makes sense, but we really don't know that much about recurrent dreams to venture a firm opinion. Here is an example of a dream that makes me believe that probably is the case. An elderly woman in therapy who had been dealing with issues of declining capabilities and increasing insecurity was very concerned about how other people might react to her. In her dream she was back in the work situation where she once had been a standout and was asking her former boss and a fellow employee for a ride home. As the car approached her home she was suddenly lost. Many trees had grown around the area and she somehow missed the road that led to her house. They all had to leave the car and look for the road. She tried to reassure her former coworkers that she knew the way but she was just stumbling about. She felt they must be looking at her and wondering. The dream woke her and after a while she fell asleep once more. Sometime later in the night she had the same or a very similar dream again. When she talked about the dream, she had little hesitancy in relating it to the issues that had brought her into therapy.

The dream was unpleasant but far from what we would consider a nightmare. Research suggests that for children and young adolescents, recurrent dreams often have the flavor of a nightmare. The dreamer is often being chased by some threatening creature, perhaps an animal or a monster, perhaps something seen in the movies, or on TV, or read about and imagined from a book.

My own research on recurrent dreams suggests that while they continue into adulthood, they are less likely to be as overtly threatening as they are in childhood and young adolescence. The monsters and witches that appeared in children's dreams were less likely to be present in the recurrent dreams of adults. The threatening circumstances young adults bring up tend to be more realistic, such as being in fires or storms or floods. The dreamers often are hiding, watching or running away. Interestingly, in my research, there were a fair number of adult recurrent dreams which were dreams of falling. The *non-threatening* recurrent dreams of adults typically were descriptions of places or settings, or dreams about friends and people they know.

Not everyone experiences recurring dreams. In my studies with college students, only slightly more than half of those questioned reported having such dreams, and I think it would be the rare person who is bombarded by such dreams night after night. Interestingly, the students in my research who reported having recurrent dreams were experiencing more daily problems in their lives than the students who did not report having such dreams, and there was a gender difference. The young women in our sample were more likely to report having recurrent dreams than the young men.

6

Gender and Age
Differences in Dreams

What are some of the gender differences in the content of dreams?
As we look at gender differences in the content of dreams, we should keep in mind that women tend to be more interested in dreams, more likely to speculate about them, more likely to view dreams as outgrowths of recent experiences, more likely to recall dreams and more likely to talk about their dreams with other people. This may reflect the high interest many women have in understanding human experience, whether this pertains to their families and close relationships or to their own inner selves.

The most extensive and reliable information about gender differences in the content of dreams comes from studies which are clearly dated. I was unhappy with this, so I looked through the recent online abstracts of the National Library of Medicine to see what might be new about gender differences in dreams, and with only a few exceptions, I found little new information. So, I must reluctantly depend to a large extent on research findings which were valid in their day, but could be misleading in today's rapidly changing world.

Two areas that immediately come to mind when we explore gender differences in dream content would be the presence of sexual and aggressive imagery. While in the most extensive early study (by Hall and Van de Castle), men more often reported sexual content in their dreams than women, the authors expressed skepticism about the accuracy of their data, because of the then prevailing cultural reluctance to discuss sex openly.

The much more recent Canadian study I also cited did not find gender differences in sexual dreams—and this may be surprising to many people. Have the dreams of women really changed in this respect over the years, reflecting changed attitudes toward sex in our culture, or does the newer research merely reflect an increased willingness of women to report sexual activity in their dreams that was there all along? Good question. No clear answer. Both explanations may well be true. I suspect that if questions about sexual content in dreams were posed by dream researchers 100 or more years ago, back when Sigmund Freud was developing his theories about dreams, one might well have found that the dream reports of men would have contained much more sexual activity than the dreams of women. The increased openness about sex in Western societies over these years has been dramatic.

Earlier in the book, we discussed an excellent review of research on the content of dreams conducted by William Domhoff. I'd like to cite it again when considering gender differences in aggression in dreams. First, some raw numbers. Aggression, defined as a desire or action to annoy or harm another person, occurred at least once in nearly half of the dreams considered, and there was very little difference between men and women: 47 percent for men and an almost identical 44 percent for women. In men's dreams, aggressive thoughts and acts were more repetitive than in women's dreams, and perhaps most notable, in men's dreams, aggression was more frankly physical, going beyond intentions and words, really lashing out.

Research has indicated that men tend to have substantially more male characters in their dreams than women, while women have only slightly more female characters in their dreams than men. Perhaps this difference reflects centuries-old views that may persist in many places—that in the "real world," building roads and houses, making a killing in high finance, running businesses and nations, and smashing heads and bodies on the football field, it is men who really count.

The dreams of women are somewhat more likely to include expressions of emotion. There is also some evidence suggesting that women's dreams contain more family members, babies and children. There is also a tendency for the dreams of women to occur in an indoor setting.

When you put these findings together, the pattern that suggests itself is that to a significant extent, women's dreams tend to reflect the traditional roles that women find themselves in. The problem with this conclusion is that most of these studies are dated. Some of the research that is usually cited was carried out a generation or two ago. If it is true that women's dreams tend to reflect women's roles in our society, then the content of women's dreams may well be altering, for it is abundantly clear that women's roles have changed dramatically in recent years. One look at the numbers of men and women in the classrooms of today's law schools and medical schools makes this point succinctly. Women are now doing many of the same things that men used to do more or less exclusively outside of the home. I suspect that some of these gender differences in the content of dreams would still be true today, such as the incidence of *physical* aggression, but some of the other differences might be disappearing.

What are age differences like in dream content? What are the dreams of children like?

The answer to this question is more puzzling than you would think. There appears to be some differences in the dreams elicited from children in the sleep laboratory and at home. Some researchers have reported that when children were awakened in a sleep laboratory, their dreams seemed to have a static quality. The child often reported static pictures somewhat analogous to what you might see in a photograph album. Frequently, the child did not see himself or herself as an active participant in the dream and there was also little activity among the other characters in the dream. The child reported a relatively large number of animals as characters in the dream.

Studies carried out in the home reveal a very different picture of children's dreams. When children have been asked about their dreams after waking in the morning, we find a very different kind of dream report. Children as young as preschoolers present themselves as playing an active role in their dreams. These dream reports contain more human than animal characters and social interactions among these people are typically present. It is hard to interpret these different observations from the sleep laboratory and the home, but there may be something in the laboratory setting with all of the gadgetry and the strangeness of the situation that distorts the child's report of his or her dreams.

In one home dream study, which was very carefully carried out by a team of researchers in Hungary (led by Piroska Sandor), there were many findings which should particularly interest parents of young children. A brief word about the method of data collection. The parents collected the children's dream reports shortly after the children woke up in the morning using an interview technique standardized for the study. The dreams were collected for a period of six weeks, which makes these data an unusually reliable and representative sample of the children's dreams.

The dream reports for the youngest children tended to be very short, averaging not more than 30 words. The dream reports for the slightly older children were a little longer, averaging around 50 words. Human characters, not animals, predominated the dreams, and many of the characters were, not surprisingly, members of the child's family.

What were the dreams like? Here are a few examples from the dream reports that were included in the report of the study.

A four-year-old boy dreamed that he ate some cookies and then went to the playground at the shopping mall.

Another four-year-old boy reported a dream about being on a ship that started to sink. His father and brother swam over to him and then deep-sea divers rescued them, taking them to dry land.

A four-year-old girl dreamed that she took a hammer from another little girl who started crying.

A three-year-old girl dreamed that she was run over by a car.

A seven-year-old girl reported a dream in which she and a friend escaped from school, climbed over the fence and went to an amusement park.

A five-year-old girl dreamed that her mother was telling her a good night tale.

A five-year-old girl dreamed that she went to the city park with her father, mother, and her siblings, Lily and Bende. They went for a walk and arrived at a garage.

A five-year-old girl reported a dream in which a tooth fell out of her mouth into her hand. She told her teacher about it but the teacher did not know where to put the tooth. She also said (this could be in another dream; I'm not sure) that a bad person came into her house and pretended to be her mother ... and she really thought she was her real mother.

These snippets seem consistent with the statistics the researchers presented when analyzing their data. Both the dream fragments and the statistics suggest that these children's dreams were anything but passive; the children were very much involved and were almost always part of the action. The characters in the dreams were largely family members and the settings seemed, for the most part, to be friendly places.

There was, however, some negativity in the dreams. For example, the dream in which a girl took a hammer from another child who cried and instances of anxiety dreams of nightmarish quality such as the dream in which family members had to be rescued while in the water and, of course, being run over by a car.

The researchers reported that in more than three-fourths of the dreams, the dreamer did appear in the dream in an active role and that 90 percent of the dreams contained at least one activity. More than half of the dreams were positive, and about one quarter negative.

Research reported by other investigators indicates that scary dreams are fairly common among very young children. Such dreams tend to be more prominent in somewhat older children, and then begin to decrease in the preteen years. As I noted in my research on recurrent dreams, scary dreams about imaginary creatures tend to decrease as the child matures.

How about the dreams of the elderly?

With the caveat that I am not satisfied with the amount of research that has been undertaken to study this question, it appears that sometime during middle age and in the years beyond, the dreams of older people begin to exhibit changes. While older people may report dreaming as frequently as ever, there appear to be changes in the content of their dreams. In general, there is less explicit emotion in the dreams, and less aggression in particular. One might speculate that the dreams of older people reflect a more stable pattern of living and a more stable outlook on life. A study of older women suggests that there is less emphasis on the dreamers themselves playing a pivotal role in the dreams. The researchers talked about this as a tendency to narrow down internal personal investments. If this interpretation is correct, it may be that dreams at this stage of life reflect a real-life decrease in what was in years past a multi-faceted responsibility. Researchers have also reported a decrease in the number of familiar characters in dreams in women as they grow older. This, too, may reflect real-life decreases in interpersonal responsibilities and involvement.

Are there any studies of the dreams of people who are nearing the end of their lives?

Patients in hospices have been asked about their dreams. The researchers reported that the patients felt their dream experiences were very real. Many of the dreams were about people they had known who had died. There were also dreams about the friends and relatives they would leave behind.

7

Bizarre Dreams and Dreams and Mental Illness

We see that many dreams reflect our experiences and how it might be possible to trace these experiences to learn more about unresolved concerns. This makes sense for the relatively tranquil, mundane dreams we often have, but what about the bizarre dreams we sometimes experience? Is it always possible to link such dreams to the experiences we have had in our waking lives?

I doubt it. The child who dreams of a monster chasing him or her has never encountered the monster. The adult who has a bizarre dream that takes him or her to strange and exotic places where improbable things happen is unlikely to be drawing on direct experience. I think it would be pushing things very hard to try to make a connection between some of our more exotic dreams and our real-life experiences. One might offer an explanation that, in this day and age, we are constantly exposed to indirect experiences of a very visual nature in our high-definition televisions which give us very clear pictures of what is happening in the far corners of the earth. We see all too graphically the effects of natural disasters such as earthquakes and tornadoes as well as the effects of human-engineered violence such as war. Research suggests that such exposure may well affect our dreams. At the same time, we know that in the pre–Internet, pre-television, even pre-radio times of not so long ago, people had bizarre dreams, so something else is going on, and the explanation may well lie in the activity of the brain.

Some researchers have looked for a clue to understanding bizarre dreams in schizophrenia. They see parallels in the waking thought and imagery of some schizophrenics and the sometimes-bizarre content of dreams. Researchers have been studying these correspondences both on the experiential and neurobiological levels. It has been suggested that there is a similarity in the state of the prefrontal cortex during REM sleep and schizophrenia.

We have previously mentioned some of the ideas of the developers of the activation-synthesis hypothesis of dreaming. Remember that they viewed dreaming sleep as largely physiologically determined. They have drawn on this line of thinking to offer an explanation for bizarre dreams, which they defined as an impossibility or an improbability of the plot and other features such as thoughts and emotions. They suggest that dream bizarreness represents a confusion of thought processes probably rooted in the neurophysiology of REM sleep. These theorists point out that in our waking states, our brains are usually capable of remaining focused on the normal flow of incoming information, and we are thus able to orient ourselves and make sense as to what is going on, placing what we are attending to in a sensible sequence. During REM sleep, however, this ability breaks down and disparate elements of consciousness are introduced into the brain which can no longer sort things out and becomes disoriented.

How do researchers assess the level of bizarreness in dreams?

Some investigators have attempted to quantify bizarreness by examining the plot, the characters, the actions and the thoughts of the characters, making a judgment whether these aspects of the dream are physically impossible or at least improbable. For example, the researchers would look for such things as changes of identity in the characters or shifts in time and place.

Here is an example of a dream that is bizarre by any standard. It was reported by a young woman who was a control subject in a study comparing dreams of people with schizophrenia with the dreams of those without. In the dream, the woman was in her hometown and knew she had to marry. The oddity was that the person she was supposed to marry had already married her in the past but now it was her time to marry him. (You can see the oddity already.) Anyway, as the dream progressed, she went to the church where there were many guests and she was dressed in white and waited at the altar for him. He didn't arrive and she became angry. She sat down among the guests and suddenly a group of people dressed like soldiers or, as she put it alternatively, as a marching band passed in front of her. To make the situation even more bizarre, a small open train followed the marching band and the groom was sitting on top of it, in front of a table full of food including some meat. The train stopped in front of her and he ate some of the meat. Her once and future husband said he was proud that she had waited for him and watched him eat.

This dream by a person without schizophrenia was full of oddities and incongruities. It was coded as pretty bizarre in the study. Compared to most dreams that are pretty routine, this one is full of interesting imagery, almost surrealistic. I can picture it in my own mind and find it fascinating.

Keep in mind that dreams later in the night tend to be longer, contain more emotion, and importantly, are more bizarre.

Many of us have unusual dreams from time to time, dreams with sudden events and incongruent elements like the dream just related. You said that researchers have observed similarities between dreams and the symptoms of the mental illness schizophrenia. Could you expand on that?

Yes. People have noticed the similarity for some time. Psychoanalysts Sigmund Freud and Carl Gustav Jung remarked on these similarities. For example, Jung wrote, "If we could imagine a dreamer walking around and acting his own dream as if he were awake, we would see the clinical picture of dementia praecox." Dementia praecox was an early term for a certain type of schizophrenia. Members of the international group of researchers that carried out the research I just alluded to, comparing the dreams of people with schizophrenia with those without, are among those who have pointed out similarities between the abnormal mental state of schizophrenia and the normal state of dreaming. Researchers are looking for these similarities on a number of levels including the psychological and neurochemical.

Clearly, we have a lot more to learn about the nature of bizarre dreams that occur in people without schizophrenia and what that may or may not teach us about schizophrenia. One hopes that such research may offer us insights into the disease.

As an aside, I am reminded of an experience I had a long time back, when I was in graduate school and taking a course in abnormal psychology. The textbook I was using made a strong case that psychoses such as schizophrenia were merely extensions of what we all experience in normal life. While I mastered the text for the upcoming exam, I never believed a word of what was written there. I never saw people without schizophrenia report they were hearing voices or proclaiming paranoid delusions. One day, a person who worked frequently with people with schizophrenia in a large mental hospital came to class to fill in as a substitute teacher, as our regular professor was ill. He said that what we had been reading in the textbook was incorrect, that malfunctioning of the brain had to be involved, and, of course, he was right. One of the nice things about science is that it tends to be self-correcting.

Tell us more about the dreams of people with schizophrenia?

As we have noted, the bizarre nature of dreams reported by acutely ill people with schizophrenia has been observed for many years. Some of the best illustrations of such dreams can be found in a study carried out in the early 1970s. Dreams were collected from hospitalized women with schizophrenia, most of whom were in acute conditions. The dreams of these patients were compared with those of female college students of the same age. While the dreams of the college women dealt with the ordinary experiences of life, such as dating, going to parties, and shopping, the dreams of those with schizophrenia were often dramatic, stark and sometimes grim. The researcher described the dreams of the women with schizophrenia as overwhelmingly threatening. Many of the dreams of the women with schizophrenia would easily meet the standards for bizarreness that we have mentioned. For example, aggression reported in the dreams was often life threatening and sometimes the dreamer was actually killed in the dream. Talk about improbabilities in dream actions. In one dream, a volleyball hit a girl, pulverizing her into a million pieces. In another dream, the woman was being decapitated, her body being cut in half. Interestingly, the dreams of the women with schizophrenia often portrayed the dreamer as being unable to cope with the threat. They seemed helpless. In contrast, in the dreams of the control subjects, the women tried to deal with the problems posed in the dreams.

Researchers showed pictures from a projective test, the *Thematic Apperception Test*, to a group of people with schizophrenia and asked them to make up stories based on what they saw in the pictures. The researchers rated the pictures for the bizarreness of content in their stories and did the same thing for their dreams. The results were similar. The cognitive and emotional distortions of schizophrenic experience can extend to both waking life and dreams.

What do we know about the dreams of depressed people? I would think that they would be rather short and maybe sad.

There is a problem in drawing conclusions about the dream recall and dream content of seriously depressed people, because many of these people suffer from severe insomnia, an integral part of their disorder. Chronic insomnia means less sleep and therefore less opportunity for dreaming. I have seen reports that insomnia may occur in 60 to 80 percent of patients with depression. One has to consider this fact when comparing the dreams of depressed and non-depressed controls.

While the answer to the question posed above is generally *yes*, the dreams of people with depression are often short and unpleasant, and their recall for dreams tends to be poor, the answer must be a very qualified one. I think dream recall, and the length and quality of the dream itself, is probably influenced by how sad and depressed a person is, how long these feelings have persisted, whether the disorder is bipolar, and, importantly, whether or not the person is using antidepressants.

Before we get too deeply into this subject, I think it would be well to say a few words about clinical depression. We usually think of clinical depression as a group or constellation of symptoms. The symptoms include a dysphoric mood (feeling sad or blue), loss of interest or pleasure in one's usual activities, decreased energy, feelings of worthlessness or guilt, disturbances in sleep and appetite, and suicidal thoughts. A person who is diagnosed as clinically depressed may not exhibit all of the symptoms, but if most are present and persist over time, it makes good sense for him or her to seek help, for there are effective treatments, both drug therapies and psychotherapy, which can be very helpful.

Let's talk about dysphoric mood for a moment, one of the benchmark symptoms of depression. Most of us experience sad and blue moods from time to time. Things go wrong and we react, but, unlike people who are clinically depressed, our moods don't stay that way very long. We feel better and our zest for doing things returns.

Interestingly, even our variable day-to-day changes in mood can affect our ability to remember dreams. Remember our discussion of mood and dream recall? Particularly the study Roland Tanck and I carried out in which we asked students to keep a specially devised diary in which they answered questions about their moods, and in the following morning recounted any dreams they may have had during the night? We found that morning dream recall was *more* frequent following the days in which students' moods were more dysphoric, when they had felt sad and lonely.

Let's return to the dreams of people who are clinically depressed and whether their dreams tend to be short and sad.

Early studies of people who were clinically depressed (patients who were experiencing severe, prolonged depression, and often hospitalized) reported that the patients tended to have rather short dreams, but surprisingly, the emotions expressed in the dreams were not all that negative. It was if their dreams were somehow cutting them off from the emotional debacle of their daily lives. Some of my patients who were extremely depressed had very sparse dream reports, if they remembered dreaming at all.

An example of the shutting off of negative emotions in dreams was provided in a study of hospitalized patients with severe depression. One of the dreams reported by these patients was that of a man who dreamed he was visiting a coffee house in Venice, Italy, enjoying the ambience, including beautiful girls and live entertainment. During this very pleasant scene, a man announced that they would be sailing around the city in a gondola.

More recent studies have reported more negativity in the dream content of people with depression than seemed to be the case in the earlier studies such as the one above. As I suggested, the depth of depression and the availability of improved treatments may make a difference. In any event, this negativity in dream content was clearly evident in a recent study where the sample selected were people whose depressed mood was mixed with anxiety. Not only were their dreams uniformly negative during REM sleep (more negative emotions and acts of aggression compared to control subjects), but the process of REM dreaming itself made them feel worse. The representation of themselves in their dreams was much more negative after REM sleep awakenings than after NREM sleep awakenings. And consider this: REM deprivation often has a salutatory though transient effect on improving moods of patients with depression.

Investigators studying the dreams of people with depression have taken a careful look at the indicators used to assess REM sleep. People experiencing depression tended to have decreased REM sleep latencies (shorter periods of time between falling asleep and the onset of the first period of REM sleep). Dreaming would tend to come on more quickly. I am not sure why this is, but it may relate to the possibility of REM rebound that depressed people might experience because of the sleeplessness that is often part of clinical depression. Interestingly, these subjects with depression also tended to show increased REM density (the number of eye movements in a given period of time).

Many people who experience depression today take anti-depressive medications. Some of these medications, particularly the tricyclic antidepressants, tend to affect dream recall, making it less likely that dreams will be remembered. The use of tricyclic antidepressants may also affect the emotions expressed in the dreams. A research review reported that tricyclic antidepressants produced more positive emotions in dreams. A study using an SSRI type antidepressant, Escitalopram, for eight weeks also reported improvements in the quality of dreams. The findings from this study suggest that when people feel less depressed in their daily lives as a result of taking antidepressants, the content of their dreams tends to be also less negative.

All this sounds fine—an added benefit of taking antidepressants—but things could get a bit tricky. The changes in dream content when starting and stopping antidepressants are not entirely predictable. It has been reported that stopping some types of antidepressants may cause nightmares.

8

Anxiety Dreams, Nightmares and Night Terrors

People who are both anxious and depressed seem to have troubled dreams. In the discussion about emotions in dreams, you said that apprehension was the most frequent emotion observed. These observations make one wonder what the dreams of people who are experiencing high levels of anxiety are like.

When I think of the word "apprehension," I think of an expectation that something very unpleasant, harmful, or even injurious may happen (or worse, *is going* to happen), the time frame involved is frequently short-term, soon, maybe even immediately, and the expectation is often accompanied by uncomfortable, even queasy, feelings.

How does apprehension differ from anxiety? When anxiety is a short-term reaction to a perceived threat, it can be almost indistinguishable from the description I offered about apprehension. Once again, there is usually an expectation for something unpleasant to occur soon. Our minds focus on the possibility of dealing with such an event while our bodies may react by mobilizing for a response. In a word, we feel tense.

Anxiety, however, can also be long term, even chronic. It may persist for weeks or months. People may feel anxious without any specific threat in mind. They simply feel tense most of the time. For some people, the situation may reach the point where they develop an anxiety disorder. In one of these anxiety disorders, called *generalized anxiety disorder*, a person may worry a lot, feel jittery, even shaky, and experience a variety of physical symptoms such as sweating, feeling dizzy and sometimes a pounding heart.

Much of the research to date on the dreams of anxious people has been focused on the relation between measures of chronic anxiety in our everyday lives and the frequency of disturbing dreams, and there is a positive relation. People who experience higher levels of anxiety in their day-to-day lives tend to have more disturbing dreams. This pattern often begins in childhood, sometimes as early as kindergarten. Children who are more anxious in their daily lives are more likely to experience nightmares. When we study adolescents, we find that those classified as more anxious on psychological evaluations report more disturbing dreams. Studies of adult women classified as having higher levels of anxiety find that the dreams of these women contain more incidents of aggression toward the dreamer than those of less anxious women.

Is there a difference between an anxiety dream and a nightmare?

When we speak about bad dreams, and particularly those fraught with anxiety, we may find ourselves anticipating and sometimes even experiencing a negative event. The situation, unpleasant as it is, however, still seems more or less manageable. In a nightmare, we often find ourselves directly exposed to a threatening situation that appears anything but manageable, and we may react with naked fear. The level of fear may become so great that we wake from sleep and remain uncomfortable for some time afterward.

In attempting to distinguish between nightmares and simply bad dreams, one of the criteria that researchers have utilized is that in a nightmare, the individual *wakes up* during the dream. Some view waking during the threatening dream as a benchmark characteristic of nightmares. Other defining terms include *a disturbing mental experience* and, of course, *a frightening dream*. I look at nightmares as dreams so disturbing that they wake you up.

Let's look at some research. The most comprehensive study I have seen about nightmares was reported by Canadian dream researchers Geneviève Robert and Antonio Zadra. Working from their university setting, they recruited more than 500 adult volunteers. Their sample included both men and women with an average age of 32. The participants kept a written record of all the dreams they remembered for a considerable period of time, two to five consecutive weeks. From the total sample of dreams, which approached 10,000, the researchers extracted 281 dreams which were classified as nightmares. These were dreams unpleasant enough to wake the dreamer. They also collected 1,016 bad dreams, unpleasant dreams which did not wake the dreamer.

Simple mental calculations will show that most of the dreams collected were neither bad dreams nor nightmares. Something like one in ten were classified as bad dreams. Clearly, they do occur now and then, but for most of us, they are not typical. The statistics for nightmares are reassuring. Fewer than 300 nightmares out of nearly 10,000 recorded dreams is a pretty small percentage. While, as we will soon see, there are people who suffer from frequent nightmares which can wreak havoc in their lives, nightmares are pretty rare for most of us.

Here are a few more findings from the study. Compared to bad dreams, nightmares were more likely to contain dreams of physical aggression, often scenes in which the dreamer was physically threatened. Nightmares were also more emotionally intense, more bizarre, and very often had bad endings. For their part, bad dreams were more likely to involve episodes of interpersonal conflict. Unpleasant enough, to be sure, but not usually the catastrophic situations the dreamer faces in many nightmares.

I have heard that children often have nightmares and as they grow older these become less frequent. Is the frequency of nightmares related to one's age?

Yes. The experience of nightmares by children is quite common. It probably reaches a peak during elementary school years, remains strong in adolescence, and declines during adulthood. Gender differences in the frequency of nightmares—with girls reporting more of them—begin to surface in the preadolescent years. Nightmares of young children are often about being chased, usually by animals. As children grow older their nightmares become more realistic. Because children who have frequent nightmares tend to be prone to anxiety in their daily lives, it seems a fair assumption that the daytime stresses they experience play a role in their nightmares. A concrete example is the bullying some children experience as they move through the developmental years of childhood. Repeated bullying can have a significant impact on children and researchers have found that it is associated with an increased frequency of nightmares.

What is a night terror and how does it differ from a nightmare?

Dreams in which we feel apprehensive, frank anxiety dreams and frightening nightmares all tend to occur during REM sleep. *This is not usually the case for night terrors.* Night terrors usually occur during *NREM* sleep, particularly during slow wave (delta) sleep. As there is more of this type of sleep in the earlier part of a night's sleep than in the later part where REM sleep is more concentrated, we would expect night terrors to occur more frequently at this time of night than in the late hours, such as the hours before dawn, and this is the case. About the only good thing you can say about night terrors is that they afflict only a small percentage of the population, perhaps one percent of adults. Unfortunately, the percent of children affected is somewhat higher, but thankfully, children often grow out of it.

I have never seen a person wake up with a night terror, but from what I have read, the event can be very scary both to the person experiencing the night terror and to the person watching it. The person experiencing the night terror wakes up suddenly, often with eyes wide open, a look of fright on his or her face, often screaming, usually sweating, and breathing rapidly, with a very elevated pulse. He or she may be thrashing about. He or she will probably be confused, not recognizing what's going on, feeling distraught, and may be inconsolable when attempts are made to relieve the fear and panic. Interestingly, incidents of night terror may be followed by amnesia for the event the next day.

Here is a report of a repeated case of night terrors experienced by a four-year-old boy. His symptoms seem comparatively mild for this type of sleep disorder, but the case is instructive because of how young the child is.

The boy's parents noticed that for about a month, he woke up in the middle of the night. That in itself wouldn't be all that unusual, but they found him standing somewhere in the house, crying, disoriented, breathing rapidly and sweating profusely. When they tried to comfort him and get him to return to his room, he became very upset, struck at them and screamed loudly. This screaming and fighting went on for several minutes and then suddenly stopped. When he seemed calm again, they put him back in bed and he slept through the rest of the night without further incidents. In the morning, he woke up in his usual happy mood and had no recollection of what had occurred. He had experienced a night terror, sleepwalking and had no memory for it.

If a person wakes with a night terror and is terrified and confused, could he or she be in danger?

Yes. People who suffer from night terrors often experience frightening images and may react by leaving the bed and sleepwalking. The individual may be trying to escape from a perceived attack. In such circumstances, the individual, in a confused state, could end up hurting himself or herself as well as others who may try to intervene. Caution when dealing with such events is advisable and seeking professional help would be the prudent course of action.

To review about disturbing dreams for adults: for most adults, bad dreams are not the typical dream but occur fairly often. Frightening nightmares that cause the dreamer to awaken are much less frequent and usually occur during REM sleep. Night terrors are uncommon, something most adults are unlikely to ever experience, and occur during NREM sleep.

Yes. The bad dream is the most frequent of the three. An occasional nightmare is not that unusual for most adults and their adverse effects are usually transitory. Night terrors are much less common, but much more serious. However, it is important to point out that nightmares can be a real problem if they occur frequently.

What would be the criteria or measuring stick for saying a person has frequent nightmares?

Some researchers studying nightmares have used the criterion of having at least one nightmare a week. Estimates for the prevalence of people who have frequent nightmares in the population is in the five percent range for adults. As we have indicated, the prevalence of frequent nightmares for children and adolescents is higher.

What would the sleep experience be like for people who report frequent nightmares?

In a word, *lousy*. They often are afraid of going to sleep because they fear they will experience nightmares. They frequently report it's very hard for them to fall asleep again after a nightmare, and their sleep is often described as fitful. They're more likely to wake up feeling restless and uncomfortable, lacking the tonic effect that most people feel after having a restful night of sleep. They often report feeling sleepy on the day following the nightmare. Their mood can be dysphoric.

I took a careful look at the data reported in a recent study of 13 people who said they have frequent nightmares. None of the subjects experienced multiple nightmares in a single night during the study. Can you imagine what that would be like? Going to sleep would be the last thing one would want to do.

Have people who experience frequent nightmares been studied in sleep labs, and if so, do they show abnormal sleep patterns such as excessive REM sleep?

There have been a few studies of chronic nightmare sufferers in sleep labs, but the results to date have been inconsistent. Researchers have reported longer REM periods, which may make sense intuitively—more time in the dreams for mayhem and for things to go wrong. However, some interesting research on nightmare sufferers carried out in the subjects' own homes using sleep lab equipment did not find any pronounced differences between these people and a control group of people who did not frequently experience nightmares on some REM measures such as REM duration and REM density. While there are no differences on these REM measures, the nightmare sufferers still reported more disturbed sleep. The researchers suggested that while nightmare sufferers do have significant sleep impairment, this may be independent of the REM characteristics measured with sleep laboratory equipment. It's an interesting possibility, but the inconsistencies in research findings indicate that we need to know more to really understand what is going on.

Are nightmares more frequent with people who have been diagnosed with psychiatric conditions?

Yes. People with psychiatric conditions tend to have more nightmares. In some psychiatric conditions, the numbers of people who report frequent nightmares can get very high. For one of the American Psychiatric Association's Diagnostic and Statistical Manual's more fuzzy categories, *borderline personality disorder* (instability in relationships, mood, self-image), about half of the people diagnosed with this condition report frequent nightmares. In *post-traumatic stress syndrome*, nightmares are a central feature of the disorder.

Frequent nightmares can cause psychological distress in the waking lives of people who have this problem. Frequent nightmares can generate anxious and depressed states and decrease one's ability to function effectively during the day, so when we state that people with psychiatric conditions report having more frequent nightmares, we must also recognize that the nightmares themselves can contribute to psychological stress. The problem can be circular: distress during waking hours may contribute to the nightmare, and the nightmare, in turn, may contribute to the emotional stress experienced during the daytime.

So, nightmares are a central feature of post-traumatic stress disorder. What are some of the other symptoms?

Post-traumatic stress disorder (PTSD) is a reaction to experiencing terrifying events such as being in combat, living in a war zone, being the victim of criminal violence such as being mugged or raped, or being caught in a deadly storm. The primary symptom is re-experiencing the frightening event, accompanied and followed by symptoms of distress such as acute anxiety. Re-experiencing may take the form of flashbacks and frightening thoughts, painful, intrusive recollections, and recurrent nightmares. There is often a pattern of avoidance—to try to stay away from the places or things that are so frightening in one's memories. The person afflicted is often easily startled and may feel chronically on edge.

Are there any ways to diminish the frequency or intensity of nightmares? Are there any drugs one could take to get relief?

A cautiously-worded review of research recently published in the *Proceedings of the Mayo Clinic* indicated that the drug Prazosin appeared to be effective in the treatment of nightmares in patients who experience post-traumatic stress disorder. The authors of the study noted that more research is needed to study the usefulness of this drug in controlling nightmares experienced by people who do not have PTSD. The review also pointed to the need for additional clinical trials.

Are there any other drugs besides Prazosin that could be effective in the treatment of nightmares?

There are a number of other drugs (e.g., antidepressants, anti-seizure drugs, low-dose cortisol) which might be considered in the treatment of PTSD-associated nightmares, but in an article entitled "Best Practice Guide for the Treatment of Nightmare Disorder in Adults," authored by a task force of the American Academy of Sleep Medicine, the available data for assessing these drugs was described as being both low grade and sparse.

Are there non-drug approaches to reducing the frequency and intensity of nightmares?

There are preventive remedies that have been tried over the centuries, some dating back to early civilizations, and there are folk remedies that are in use today. Some of the ancient practices included turning to protective spirits and burying something under the floor accompanied by words admonishing the bad dreams to leave. In some Native American cultures, a willow hoop with woven netting was hung over a child's cradleboard to catch bad dreams. It's possible that some of these preventive actions might have a placebo-like effect, if the dreamer believed in their efficacy, but I wouldn't count on it.

In its "Best Practice Guide," the American Academy of Sleep Medicine discussed some empirically tested psychological therapies recommended for the treatment of nightmare disorders. In considering the use of these therapies, and for using medicines as well, it should be remembered that nightmare disorders are not the same thing as the occasional nightmare, which most people experience. The occasional nightmare may require about the same kind of attention as an occasional headache or cold, but when nightmares are persistent and leave one anxious and distressed, there are a number of techniques used by psychologists, and particularly by psychologists who include behavioral therapy and cognitive behavioral techniques in their treatment regimen, which the "Best Practice Guide" recommends. These include *systematic desensitization and progressive deep muscle relaxation training* for the treatment of idiopathic nightmares. *Image rehearsal therapy* seems particularly effective for the treatment of nightmare disorders.

Let's begin with a few words about systematic desensitization. I used systematic desensitization in the treatment of phobias when I was a practicing psychotherapist and also in the treatment of PTSD. The basic idea is to first teach patients deep muscle relaxation, a procedure in which the patient learns to relax various muscle groups (e.g., arms, shoulders, neck) one at a time, which eventually produces an overall feeling of relaxation. When the patient is relaxed, he or she is asked to imagine some of the scenes that he or she felt threatening. The usual practice is to begin with the scenes that are least threatening and progress gradually to scenes that are more threatening. If the patient becomes agitated by the experience, one stops and retreats in the progression of scenes. With the advent of modern technology, it is now possible to use virtual reality to present the scenes to the patient instead of relying on the patient's imagination. In the treatment of phobias, it was my experience that the patients who went through this procedure were better able to tolerate these phobic-arousing situations in real life.

Imagery rehearsal therapy is a more recently developed technique than systematic desensitization. The basic assumption when applied to dreams is that if we spend time thinking about one of our disturbing dreams during our waking hours—and if we wish to modify the dream, to change what happens to make it less disturbing—we can actually write out a new, less threatening version and then use imagery techniques to constantly rehearse it. Imagery rehearsal therapy postulates that this daytime activity can actually influence the content of the dream the next time we experience it.

In the treatment of nightmares, for example, the patient is first asked to choose a nightmare he or she has been experiencing and then to construct a new dream narrative, changing part of the nightmare so that it is not as threatening. If the nightmare is one in which you are trying to get to an important examination but can't find the building or the classroom where you are supposed to take it, you might in your waking life reframe and rehearse the dream so that you do arrive there and do very well on the test. In a dream in which you are being chased by a wild animal, you might reconstruct the dream so that you are being followed by a friendly dog like Lassie.

This is undoubtedly an oversimplification of the treatment and the reader may want to pursue the more detailed references offered at the end of the book. With this caveat, here is a case that illustrates how the technique has been used in clinical practice.

The patient, a 69-year-old male suffering from post-traumatic stress disorder after two military tours in Vietnam, was experiencing intense nightmares, about 17 per week. His nightmares were violent, fighting with animals and being in combat situations. He was asked to re-script both nightmares. For the Vietnam combat nightmare, he pictured himself still in Vietnam, but as he looked over the fields and hills around him, everything seemed very peaceful and beautiful and he felt calm. He practiced visualizing his re-scripted scenarios several times a week, and after several weeks, the number of these nightmares decreased substantially to about five per week. Did this treatment eliminate his nightmares? Clearly, no, but he felt better and was able to sleep much better.

I must admit that when I first read about the technique, I was more than a little skeptical that such an approach to changing nightmare content would actually work, but I was pleasantly surprised by the evidence indicating that for diminishing the frequency and severity of nightmares, imagery rehearsal therapy may work very well. In fact, a careful review of the research evidence indicates that the technique produces results that are similar to those attributable to the drug Prazosin. If you suffer from chronic nightmares, you may want to consult a professional who is proficient in using the technique.

Drugs and behavioral techniques sound useful for adults. What about children? And particularly young children who are experiencing nightmares? What can be done to help them?

My first thought is to follow the Hippocratic Oath—to do no harm. Occasional nightmares are not unusual for preschool and school age children and will likely diminish over time. If it becomes clear that intervention is needed, parents might try some of the things they usually do to comfort their child, such as holding and cuddling him or her. Remaining with the child for a while following the nightmare may help the child return to sleep. If the child has a favorite toy such as a stuffed animal, leaving it with the child may give him or her a sense of security, like Linus and his blue blanket in the comic strip *Peanuts*. Another possibility is to leave the bedroom door open or perhaps turn on a small light in the corner of the room. Parents are usually the best guide as to what is reassuring for their children. If nightmares increase in frequency or are causing the child increasing distress, then a consultation with the child's pediatrician is certainly in order. In addition, there are professionals who are trained to help the child deal with the anxieties of his or her young life, anxieties that may be fostering the nightmares. A child psychologist may be helpful.

Anxiety dreams, and even worse, nightmares, can leave a person upset, tense, and even depressed when he or she wakes up. One wonders whether such dreams can have any positive value for the dreamer.

That is an interesting question. It seems plausible that anxiety dreams and perhaps even occasional nightmares might have had some survival value if we look at dreams from an evolutionary perspective. One can imagine that our ancestors who were relatively defenseless against predators might have had anxiety dreams or nightmares about being pursued by wild animals. One can speculate that such dreams might have helped maintain a sense of alertness against such possibilities during waking hours. Today, if one lives in a high threat environment such as a war zone, being ever alert has survival value. However, when one lives in a more normal environment, it is hard to imagine much in the way of benefits from repeated anxiety dreams or chronic nightmares. To the contrary, such dreams would tend to make a person less effective in his or her everyday life, not more effective.

Not everyone would subscribe to this position. Certainly not an orthodox Freudian psychoanalyst. Remember that Freud thought that dreams were the guardians of sleep and that behind every dream there was a wish, usually sexual in nature. For the first idea, that dreams were the guardians of sleep, you would have to stand reality on its head to think that nightmares were the guardians of sleep. We have seen that they often wake up people with a sense of fright, even terror. Even Freud somewhat grudgingly admitted that his guardian of sleep theory was unlikely to apply to nightmares. However, he insisted that there was a wish underlying dreams, even anxiety dreams and nightmares.

This idea has always seemed convoluted to me. Freud himself admitted in his writings that, on the surface, reconciling nightmares and anxiety dreams with the idea that dreams were really wish fulfillments appeared an impossible task. However, he argued that his theory of wish fulfillment was not based on the manifest content of dreams but on the latent thoughts which he uncovered through his method of dream interpretation. He believed that these latent thoughts would reveal the wish that was the driving force behind anxiety dreams and nightmares.

Anything is possible, of course, but as I said, this seems like a very convoluted idea to me. I have never seen any serious evidence from research to support it. I prefer a more straightforward view that nightmares reflect fears, not wishes. They draw on uncomfortable, sometimes scary events lodged in a person's memory. They are replays of things (with modifications) that have happened to one in one's daily life, heard about or read about or seen on television or in movies. The only clear-cut wish that I can see that may come with them is the wish to awaken and not have another bad dream when one returns to sleep.

9

Positive Effects
from Dreams on Emotions
and Problem Solving

Let's turn away from bad dreams. Can the normal run of dreams that we experience have any positive effects on the dreamer when he or she awakens?

Yes. There is evidence that both sleep and dreams may have positive effects on a person the following day. We know that a good night's sleep itself can affect the way you feel. If you sleep poorly, you are less likely to wake up with a positive mood, ready to take on the world, and many studies have shown that sleep can have a positive effect on memory consolidation. If you learn something and then sleep on it, you are likely to have a better memory for what you learned than if you simply stayed awake.

When we talk specifically about dreams rather than sleep, there is some research suggesting that dreams may have at least two types of benefits. The first is that, nightmares aside, dreams may make you feel a little better the next day. The second is that dreams may be of some value in problem solving and creative activities.

You may be wondering if dreaming really can make you feel better the next day. Think back for a moment to the research on REM deprivation. If your REM sleep during the night is diminished, and this is likely to reduce the likelihood of interesting, exciting dreams, during the next day, you are likely to feel edgy, anxious, perhaps even a little confused, and your skill in interpersonal relationships may be diminished as well. More recent research has demonstrated that REM sleep is associated with the overnight dissipation of amygdala activity in the brain (which was responding to prior emotional experiences), thereby reducing the intensity of emotions one feels the next day.

When it comes to assessing the effects of dreaming on problem solving and creativity, it is a little difficult to disentangle the effects that might ensue from dreaming from those you might obtain from simply absenting yourself from the problem—which in itself can sometimes lead to ideas which are novel and helpful. Have you ever found yourself trying to solve a problem and not getting anywhere, becoming frustrated, and simply giving up for a while and turning your mind to something else? Later in the day, or perhaps in the evening, a new idea about dealing with the problem suddenly pops into your mind, seemingly out of the blue—an idea which points you in a new direction in approaching the problem. The interval of time in which you have seemingly left the problem behind until the moment the idea suddenly pops into your head has been described as a period of incubation.

One of the best-known examples of the period of incubation happened to the distinguished mathematician Jules Henri Poincaré. Poincaré left his home in Caen, Normandy, to go on a geologic excursion under the auspices of the School of Mines. Traveling made him forget his mathematical work. When he reached Coutances, also in Normandy, he boarded a bus to continue his travel. When he put his foot on the step, an important mathematical idea came to him that had nothing to do with his previous thoughts. (For those who are students of mathematics, the idea was the realization that the transformations he had used to define the Fuchsian functions were identical with those of non–Euclidean geometry.) While he did not have time to verify the idea, he felt certain that the idea that had suddenly occurred to him as he entered the vehicle was correct.

Now, let us return to sleep and dreams. In helping you solve a problem or acting creatively, sleep has at least three things going for it:

1. It is restorative.
2. It provides a wonderful period for incubation. You are away from just about everything.

3. Brain activity is taking place, both during REM and NREM sleep, including both thought processes and imagery, and it is usually out of your control, permitting new combinations of ideas to emerge.

Let's take a look at some of the evidence for the benefits of sleep and dreams that I have alluded to. First, let's consider some research demonstrating the positive effects of sleep on problem solving. If you are given a task to solve before you sleep and then sleep on it, research suggests that you may do better afterward in solving the problem than if you simply stayed awake. Here are two brief examples of studies.

In one study, people were asked to solve a videogame problem without much success and then asked to take a nap. Subjects who took the nap were almost twice as likely to solve the problem afterward when compared with subjects in a control group that remained awake.

In a second, somewhat similar experiment, subjects were trained to perform a virtual navigation task and retested five hours after the initial training. The subjects who took a nap between the trials improved their performance more than the subjects who remained awake. Importantly, improvement was strongly related to the napping subjects reporting NREM dream imagery that was related to the virtual navigation task.

So dreaming looks like it could offer an assist to sleep in problem solving. If this is the case, here is a question that comes quickly to mind: is it the more dream-like REM sleep that is the source of inventiveness or is it the more thought-like NREM sleep?

At the present time, the answer is not altogether clear. The research on the videogame problem suggested that it was slow wave sleep rather than rapid eye movement sleep that was responsible for the improvement. All of the subjects who experienced slow wave sleep solved the problem while REM sleep did not seem to be involved. Moreover, in the virtual navigation task study, the subjects who reported that they experienced task-related mentation had these dreams not that long after sleep had begun, before the onset of REM sleep.

Reports from other studies, however, suggest that REM sleep may have a role in creative problem solving, so the question needs more investigation. Creative problem solving or simply creativity often involves the forming of old ideas into new combinations, and I suspect that both types of sleep might have a role in that process. In our own nightly sleep, we cannot distinguish with certainty which type of sleep was occurring when creative ideas occurred, so when we experience them, simply be pleased if they prove useful.

As we have said, sleep certainly helps memory consolidation and dreaming may have a role in creative behaviors, not only in everyday life, but in the work of creative artists, writers and scientists.

In the former case, there is an interesting study of college students who were asked to pay attention to their dreams, particularly those dreams in which they could identify problems they were facing in their lives. In most of these dreams the students found solutions to these problems, more so if the problems portrayed were personal rather than academic. There is at least the suggestion here that paying attention to dreams could be helpful for people in addressing issues in their lives.

In the latter case, there are some notable examples where dreams were said to have played a role in fostering creativity. I should caution that these examples provide only anecdotal evidence, not the rigor of a scientific study. Still, these reports are very interesting.

Three often cited examples are the writing of the poem "Kubla Kahn" by Samuel Taylor Coleridge, inspiration for the story *The Strange Case of Dr. Jekyll and Mr. Hyde* by Robert Louis Stevenson, and the work of Nobel laureate Otto Loewi on the chemical transmission of nerve impulses.

The poem "Kubla Kahn" is remarkable for its use of imagery. Here are the opening lines:

> In Xanadu did Kubla Khan
> A stately pleasure-dome decree:
> Where Alph, the sacred river, ran
> Through caverns measureless to man
> Down to a sunless sea.
> So twice five miles of fertile ground
> With walls and towers were girdled round;

And there were gardens bright with sinuous rills,
Where blossomed many an incense-bearing tree;
And here were forests ancient as the hills,
Enfolding sunny spots of greenery.

The poem was written in the summer of 1797 when Coleridge was living in a lonely farmhouse and not feeling well. He had taken a prescribed drug and was sitting in a chair reading a note in a book about Kubla Khan, which read, "Here the Kahn Kubla commanded a palace to be built, and a stately garden thereunto. And thus ten miles of fertile ground were inclosed with a wall." While reading, Coleridge fell into a profound sleep in which he reported that he had composed 200 to 300 lines of poetry. He said that the images rose up before him without any effort on his part with a parallel production of the corresponding expressions. (I assume he was referring to the actual lines of poetry.) When he awakened, he immediately began to write down the lines of the poem. In reading Coleridge's account, one gets the feeling that the experience was something akin to the poem being dictated to him in his sleep. In any event, as he was writing down the poem, he was interrupted by a man calling about business, and when the interview had concluded, most of his vision from the dream had slipped away, vanishing from memory, so the poem remains the fragment we know today.

Both Fanny Stevenson, the author's wife, and Lloyd Osborne, his stepson, were present when famed 19th-century Scottish writer Robert Louis Stevenson wrote his eerie tale of the kindly physician, Dr. Jekyll, who found a way of transforming himself into the vile Mr. Hyde. The novel caught the public's imagination and solidified Stevenson's reputation as one of the great storytellers of 19th-century English literature. You can find the account about how the story came to be in one of the biographies of Robert Louis Stevenson, but for the sheer pleasure of reading, I recommend Nancy Horan's engaging novel *Under the Wide and Starry Sky*, which tells the love story of Stevenson and his wife Fanny. As the story is told, Fanny was awakened in the night by her husband's horrific cries. She thought that he had been experiencing a nightmare and awakened him. Instead of reacting gratefully, he was angry, protesting, asking why she had disturbed him. He had been in the midst of dreaming—dreaming scenes of Jekyll and Hyde, perhaps nightmarish, but a riveting tale that he commenced to write almost immediately. Even though Stevenson was ill, he wrote with great speed and endurance and finished the novel in days.

Otto Loewi's story is an interesting one in itself. Loewi was born in Frankfurt, Germany, in 1873 to a Jewish family. He took a medical degree and then decided not to become a practicing physician, but rather to carry out research in basic medical science. He became interested in the problem of how signals were transmitted from the synapse of one neuron to the synapse of another. At the time, it was uncertain whether there were chemicals involved in the transmission of nerve impulses. Loewi believed that chemicals were involved but he didn't know how to demonstrate this experimentally. In 1921, he figured out how to conduct an experiment that would show whether chemicals were involved. The experiment was a success—he proved that chemicals were indeed involved in the transmission of nerve impulses, and this led to subsequent research and our current understanding of the role of neurotransmitters. Loewi stated that the idea for this key experiment came to him while he was sleeping. Here is his own account:

> The night before Easter Sunday of that year I awoke, turned on the light, and jotted down a few notes on a tiny slip of paper. Then I fell asleep again. It occurred to me at 6 o'clock in the morning that during the night I had written down something most important, but I was unable to decipher the scrawl. The next night, at 3 o'clock, the idea returned. It was the design of an experiment to determine whether or not the hypothesis of chemical transmission that I had uttered 17 years ago was correct. I got up immediately, went to the laboratory, and performed a single experiment on a frog's heart according to the nocturnal design.

In 1936, he was awarded the Nobel Prize for his discoveries. Not long after, the German army moved into Austria and Loewi was immediately arrested along with two of his children. Loewi was allowed to leave the country after he relinquished all of his possessions to the Nazis.

There are other reports of creative responses to dreams by gifted scientists, authors, and composers. Can dreams be sources of creative activities for us lesser mortals? If you have ever had a dream and followed it up with a creative activity, I would like to hear about it. Please let me know.

10

Message and Predictive Dreams

Some people believe that dreams can be messages sent to the dreamer and that dreams themselves can be prophetic—predictions or even warning of things to come. What can scientists tell us about this?

I'm afraid that science can tell us very little about either of these beliefs, but this is a good place to pause in our discussion, for these questions offer a convenient bridge to take us from subjects where science can provide a good deal of reliable information about dreams to a more problematic, murkier area—the question of dream interpretation, an area of inquiry in which the standards of science often give way to supposition, unproven theories, and too often, to extravagant claims. Before we cross this bridge, however, let's take a look at these two questions, the idea that dreams can be messages and the idea that dreams can be predictive of future events.

First, it is important to recognize that these ideas are very old, going back to the dawn of recorded history, to the papyruses of ancient Egypt and the baked clay tablets recovered from the ruins of the ancient cities in Sumer and Akkad that once flourished in long-ago Mesopotamia and the larger states such as Babylonia and Assyria that followed them. The tablets reveal a great deal about daily life in these ancient cities and on occasion give us some ideas about how these early inhabitants of civilization thought about and interpreted their dreams. The best source of information for this that I know of is a monograph by A. Leo Oppenheim entitled *The Interpretation of Dreams in the Ancient Near East*. It is a fine example of scholarship by an expert in the field. If you are interested in reading it, I must caution you that it is not easy to locate, and if you can find it, it is part of a very heavy volume of scholarly papers, not something you can easily move around.

How did the people of 3,000 or 4,000 years ago think about their dreams? Oppenheim tells us that the Mesopotamian records suggest that they wrote about several types of dreams. Two of these types of dreams are especially relevant to our inquiry here: message dreams and dreams that predicted future events. A third type of dream was what he called *mantic dreams.*

The message dreams recovered from these Mesopotamian cities are similar in some ways to both the familiar message dreams one reads about in the Bible and the dreams we read about in Homer's epics *The Iliad* and *The Odyssey* in that the messages do not come from other people, but from one of the gods in the Mesopotamian and Greek sources and from God in the Judeo–Christian religious texts. The fact that there are similarities between the Mesopotamian records and the stories from the Hebrew Bible should not be surprising, as the traditional birthplace ascribed to Abraham was Ur, one of the leading cities of Sumer. One may also find similarities in the ancient Babylonian legal codes and the ideas of justice in the Hebrew Bible. When we look at the Mesopotamian message dreams that have come down to us, the gods usually speak to only persons of high rank such as kings, not to the common people, and as one might suspect in these ancient cultures, the recipients of these message dreams were almost always men. Usually in the dreams, the god appeared near the head of the dreamer and conveyed his message. It was seldom a dialogue, more likely instructions or recommendations, and then the dreamer would wake up abruptly.

Sometimes these instructions would be of a political nature. A would-be aspirant to the throne of Assyria was told to give up the idea and support another person who would eventually become king. In another instance, the female goddess Ishtar told the dreamer to back an insurgent aspirant to the throne.

Here is an example of a message dream included in Oppenheim's monograph. Although it is somewhat atypical, I selected it because it seemed easily comprehensible to a modern audience. As Hammurabi, the king of Babylon, was mentioned in Oppenheim's introduction to the dream, it would appear that this dream report is very old indeed, recorded around 4,000 years ago.

A minor official in a small kingdom eventually destroyed by Hammurabi in his drive to extend the territory of Babylon reported a dream that he was on his way to the kingdom's capital when he stopped to visit the temple of the god Dagan. In the temple, he prostrated himself before the image of Dagan (I assume this was an idol) and immediately heard a voice which he identified as the voice of Dagan. The god gave him a message to take to the king. In the message, Dagan said he had not received information about a possible peace agreement between warring factions in the kingdom and the king's army. Dagan told the official to tell the king to send his emissaries to the temple (presumably for a briefing for the god to tell him what was going on in these negotiations). In the message for the king, he stated that he, the god Dagan, would have long ago delivered the sheiks of the tribes (the rebellious faction) into the hands of the king. It is interesting that Dagan seemed to be anything but omniscient, not even being aware of what was going on in the small kingdom, and a bit petulant as well. I can't help thinking of some of the Greek gods who at times could be quite angry and peevish, traits that may have reflected the emotions of the people they were supposed to rule.

Let us leave these dream messages from ancient gods to kings to the historians and consider another twist to message dreams—the use of dreams for communications from one person to another. You might be thinking that all I am referring to is the common situation in which a person, perhaps your best friend, says, "I had the most interesting dream last night," and tells you about it and then you and your friend discuss it, but I have nothing so mundane in mind. Rather, we will take an ever so brief glimpse into the paranormal, where some believe that dreams may be a means for transporting ideas, even messages from one person another. In doing this, we will look into some interesting research into the possibility of dream telepathy.

If I heard that someone had claimed to be receiving messages in his dreams, my skepticism antenna would turn on and the meter would immediately rise to high levels. The first question that would occur to me is the same as would occur to many people, including many readers of this book. To put the question in the vernacular, "Is this guy some kind of nut?" To phrase the question in less pejorative psychiatric language, "Is the person making this claim afflicted with schizophrenia, an all too common mental disease in which hearing voices is a prime symptom?" A fair question to ask. If the answer is no, and the individual is mentally healthy, then the claim becomes much more interesting. The next objection that would vibrate along my skepticism antenna would be the fact that conversation is a normal, not highly unusual occurrence in dreams, and therefore, no interpretation of receiving a message is necessary to explain the phenomena. An urging from someone in the dream conversation to take some course of action or even to feel some particular way might well be emanating from one's own dream construction, not from someone else's messaging. If, as I believe, many dreams have their origin in unresolved problems, then we need look no further for the source of the message than to yourself. It may simply be a reminder of a sort to oneself to do something about a problem you have let simmer.

If the claim of receiving a message still survives these critiques, I would imagine the next thing to do is to talk to the supposed messenger. If the message is thought to come from a definable person, we have an obvious path open to us. Talk to the messenger. Did the person have a dream in which he or she was attempting to send the message? That would really be something, wouldn't it? If not, we could still ask whether the sender had been trying to communicate a message to the dreamer using any means. Or, even more tenuous, was the person even thinking about the issues the receiver claims to have picked up in his or her dream? If the answer is *yes* to these later possibilities, then the situation may still be worth pursuing. One would start by looking for reasonable, non-telepathic explanations, and prominent among them would be that old bugaboo of fans of the paranormal, *coincidence.*

With the number of dreams experienced by human beings each night numbering in the billions, and the number of people who do things which might appear to act in accordance with someone else's dream also being a very large number, the odds of coincidence weigh heavily on those who jump to the conclusion that telepathic messages may be transmitted in dreams. Having said this, there are a number of such correspondences between the dreams of one person and the dreams or thoughts during sleep of another that seem downright eerie.

J.B. Rhine, best known for his dogged research exploring the possibility of extrasensory perception, collected some telepathic-appearing dreams that are real mind-bogglers. For example, during World War II, a mother dreamed that her son, sleeping in a tent on a remote Pacific island, was in imminent danger because a nearby tree was about to fall and crush the tent. She called his name out frantically as if sending a warning. While sleeping, the son thought he heard his mother's voice. Wondering where the voice was coming from, he left the tent and then turned around and saw a tree crashing behind him that actually demolished the cot he had been sleeping on. Weird, eh? Makes one think of a story Edgar Allan Poe might have written.

I suspect that the last thing an academic researcher who wants to build a reputation in the psychological sciences would want to do would be to have his name attached to paranormal research—with its far-out reputation—so I have nothing but praise for Montague Ullman, who seriously studied the possibility of dream telepathy. In brief, what he did was have two people, a sender and a receiver, stationed in rooms 90 feet apart. One of these persons, the receiver, slept and was awakened during REM periods and had his or her dream recorded. The other person, the sender, was given a reproduction of a painting to look at and instructed to try his best to communicate the image of the painting by thought alone (telepathy) to the sleeping person. The dream reports of the receiver were then given to several judges to read along with a number of pictures, one of which was the picture the sender saw. The judges were asked to rank the pictures they were given in terms of how nearly each resembled the dream report. The judges found this task difficult, but when they had made their rankings, it was possible to see if there was any tendency for the picture the sender actually used to be ranked high by the judges as suggestive of the dream report. For those rooting for a telepathic effect, the results were disappointing. Results were strictly chance. There was no indication of common features between the picture used and the dream report. The procedure was repeated a number of times with different subjects with the same negative results.

There was, however, a glimmer of hope in a follow-up study in which the researchers picked their *very best subject*, one who might just be potentially sensitive to telepathic communications, and they did find a positive result. The correspondences between the content of the painting and the content of his dreams were really quite interesting, but when the researchers tried to repeat the experiment with another person who looked like he might also be a sensitive subject, the results were chance. It was a now you see it, now you don't phenomenon, with now you don't being the usual result.

While writing this book, I searched through the Library of Medicine's abstracts for "telepathic dreams" to see if there had been more of this kind of research since Ullman's pioneering studies. I found next to nothing, only a psychoanalyst's report that one of his patients may have had a telepathic dream. Interesting, but a long way from what we would call scientific research.

Let's turn our attention now to predictive dreams. These are dreams (typically requiring some interpretation) that can supposedly predict future events. Remember the dream from the ancient Egyptian papyrus in which it was said that a dream of eating crocodile flesh indicated that the dreamer would become a village official? Dreams with a prediction attached were very common in a dream book found at the court of the Assyrian king Ashurbanipal. (You can see photographs of the clay tablets on which the dictionary was inscribed in Oppenheim's monograph.) This was one of the first known examples of dream dictionaries, which we shall discuss in more detail later in this book.

Let's look at a few examples from this roughly 3,000-year-old Assyrian dream dictionary. Sticking with the idea of eating, here are interpretations for dreams of eating the meat of a wild bull, not a very common dish these days. The prediction? The dreamer's days will be long indeed. But wait! Here's another prediction. Dreaming of eating a wild animal will be followed by deaths in the family. Sound a bit contradictory, trying to have it both ways? Perhaps not. The dictionary writer may have been thinking of a long life for the reader, and the opposite for someone else. Maybe a bad-tempered, miserly uncle. Here's another one. Dreaming of eating the innards of an animal predicts peace of mind. (I think I'd rather see a shrink.) There are even interpretations of cannibalistic dreams, including eating different parts of the body, but I'd rather not go into that at all.

In the Assyrian dream dictionary, there were predictive consequences for dreams in which one carried out everyday activities. For example, if a man dreams of hunting in the desert, he will become sad. If he dreams of going to a vegetable garden, his work will become harder. There are predictions for dreams in which people are given gifts. For example, if someone is given a wagon, he will obtain his desire, but if the gift is a door, the dreamer will grow old. (That's one door prize you wouldn't want.) If you dreamed of visiting certain cities (e.g., Girsu), you would experience joy, but if you decided to visit Lagash, you would be robbed.

The list of possible dream events goes on with the consequences spelled out, but I think that will do for now.

You can see that some of the predictions made from dreams in the Assyrian dream book were cut and dried—i.e., if you dreamed about such and such, the following things would happen to you or to others. If a citizen of this ancient civilization believed what was written in the book was credible, and happened to dream about a particular activity included in it, well, he might believe he knew what life held for him. Believing this, he might expect the prediction to come true. There would always be the possibility of a self-fulfilling prophecy to help things along. Encouraged by a favorable prediction, the dreamer might do things that might make the prediction more likely to occur in reality, or if the prediction was truly ominous and the dreamer thought that the predicted misfortune was somehow preventable, he or she might take steps to try to avoid the event.

A classic tale of taking steps to prevent the fulfillment of an ominous prediction was recounted in Herodotus' book *The Persian Wars*. Herodotus, often referred to as the "Father of History," traveled widely in the lands of the eastern Mediterranean, recording his observations of the people and the lands he visited and recounting the stories he heard. This particular story concerned Astyages, the king of the Medes (the Medes lived among the Persians and at that time dominated them), and his grandson, Cyrus, who became the ruler of the mighty Persian Empire that stretched over much of the Ancient Near East and for decades threatened to engulf Greece as well.

Astyages had incredible dreams about his daughter, Mandane. He asked the Magi who were thought to be skilled in dream interpretation to tell him what the dreams meant. They told him that the dreams foresaw that his daughter would have a son who would supplant Astyages, taking his throne away from him. Frightened by this dream, the king instructed one of his trusted underlings, a man named Harpagus, to steal his daughter's newborn infant and dispose of it. Harpagus was reluctant to do this himself and gave the child to a herdsman to do the deed. When the herdsman and his wife saw the child, instead of killing him, they decided to raise him themselves. In a very long tale, Herodotus related how the boy grew to manhood, discovered his true identity, rallied the Persians against their masters, overthrew Astyages and became the ruler of the Persian Empire.

I have no idea whether Herodotus believed this tale to be fact or fiction, or perhaps a bit of both, but you may notice a slight similarity with the biblical story of the infant Moses being left in the bulrushes or a very pronounced similarity with Sophocles' play *Oedipus Rex*. Like Cyrus, Oedipus was left to be disposed of by his father, the king of Thebes, after the king was given one of those famous doomsday prophecies about his son, and like the story about Cyrus, the death sentence was not carried out. Oedipus was raised by a shepherd, and when grown to manhood, he unknowingly killed his father, married his mother, blinded himself, and then suffered the final indignity: he became the poster child for a Freudian complex.

The narrative of unsavory events in the past coming slowly to light in a process of discovery to impinge on the lives of those who follow is a trusted plot line that has more than once found its way into contemporary fiction. I must plead guilty of doing so myself in my romantic suspense novel, *Look into the Past, Mercedes, If You Dare!*

I have always been skeptical of the idea of predictive dreams. One of my reasons is that dreams are largely based on autobiographical memories and many of these memories concern behaviors, thoughts and conversations that a person once had and is very likely to repeat in similar ways in the future *whether or not they appear in the dream*. Therefore, what you may think is a predictive dream may be simply something you would have experienced whether you dreamed or not.

About 100 years ago, Freud wrote what he thought about the possibility of prophetic dreams, summarily rejecting the idea. Freud argued that when one takes into account the untrustworthiness and unconvincingness of most of these reports, the possibility of falsification of memories, and the ever-present role of chance, there would be little left to account for in dismissing the likelihood of prophetic dreams. While I often take exception to Freud's ideas, I couldn't agree more with him on this one.

I think we have walked over that bridge between what science tells us about dreams and the attempts to understand what dreams might be telling us about ourselves.

PART II

*What Dreams May Tell Us
About Ourselves*

11

Comparing Your Dreams
with the Dreams of Others

In Part I, we took a broad look at what scientific research has revealed about dreams. We discussed the discovery of REM sleep and how it was linked to dreaming. We considered the differences between the dreams that occur during REM sleep and NREM sleep and how these different types of sleep were related to the activation and deactivation of different parts of the brain. We discussed many questions that researchers have investigated about dreams, such as factors influencing dream recall, gender and age differences in dreams, lucid dreaming, the dreams of mentally ill people, nightmares and night terrors, and the possible value of dreams in problem solving and creativity.

Having concluded our survey of what science has told us about dreams, we now turn to the ways people have tried to interpret their dreams. We know that people have tried to interpret their dreams since the beginning of written language and we can assume that they have tried to do this for centuries, if not millennia, before they were able to record their ideas for themselves and posterity. Judging from our initial examination of theories of dreams, we can already see that there is not one agreed-upon way to interpret dreams; rather, the approaches are varied.

Before we begin our examination of these different methods of dream analysis, I would like to take a preliminary step. Because many readers are likely to think about their own dreams as they read about these approaches to interpretation, I think it is a good plan to begin by presenting some statistical information that will enable these readers to have a more precise understanding of what the dreams of other people are actually like, so they can see where their own dreams are similar to most people's and where they are not.

After presenting this information, my plan is to describe several important approaches to dream interpretation, looking at them critically, and whenever possible, use a set of common benchmarks to compare and contrast them.

Here is the road map for this part of the book.

One. Comparing one's own dreams with those of others.

In this preliminary step, I will try to assist you in assessing how similar your own dreams are to those of other people. Before asking what your dreams might mean, it seems useful to inquire whether your dreams are similar to those of other people or whether they tend to be different. Do you dream about the same things that other people do and do you experience similar emotions?

In looking at this question, we will draw on the data that researchers have given us about the observable content of dreams. These are detailed analyses of what people report in their dreams, examining both the content of the dreams, and the emotions experienced by the characters in the dream. Using a large sample of dream reports, researchers have made it possible to estimate the frequency with which various settings, objects, actions and emotions occur in dreams and in so doing to define the common features of dreams. This information should prove useful in gauging whether one's own dreams tend to be typical or unusual, and if unusual, in what ways.

Two. Interpreting dreams using the psychoanalytic methods of Sigmund Freud.

We have discussed Freud's theories briefly, but now we shall see how Freud went about using the technique he developed, *free association*, to dreams, as his basic tool to try to unravel the meaning of dreams.

Three. Examining the dream analysis method of Carl Jung.

Jung's ideas are important, and we shall see how he went about interpreting dreams.

Four. Looking at the oldest method of dream analysis: the dream dictionary approach.

In this approach, dating back to the ancient world, people have selected settings, objects, and events in dreams that seem to be typical and have offered specific meanings for them. Our focus will be on contemporary versions of this approach, books containing dream dictionaries taken from the shelves of our public libraries.

Five. Looking at the meaning of dream-related incidents using the Dream Incident Technique.

In exploring this approach, we will highlight the Dream Incident Technique, a research instrument developed using the tools of contemporary psychology. It provides a more objective, scientific approach to dream analysis.

We will conclude with some suggestions that you may wish to consider in making sense of your own dreams.

In describing each method of dream analysis, we will examine the way these approaches view dreams, the assumptions that are made, and the methods used in analyzing or interpreting dreams. To make these methods explicit and clear, we will present examples of dreams and the way they are analyzed. Then we will consider the degree to which these approaches seem consistent or inconsistent with the scientific knowledge that has been compiled about dreams, drawing on our review in Part I. As our most critical benchmark, we will ask what evidence there is for the validity of these approaches to dream analysis.

In exploring this question, we will inquire whether predictions can be made from dream analyses that can be objectively verified. If so, we will question whether these predictions been have born out and, importantly, reported in a way that the results can be critically assessed by others.

Comparing your dreams with the dreams of others.

If you are interested in comparing your dreams with the dreams of others and have good recall of your dreams, make an effort to record your dreams by keeping a pen and notebook by your bed or, even better, if you have one, a recording device. In the bleary moments of awakening, turn your dream recall into written words as soon as you can. Dreams that are not vivid, meaningful, frightening or remarkable in some way can vanish like a will-o'-the-wisp, so the proverbial admonition "make haste" certainly applies. When you have collected a sample of dreams, I would suggest a minimum of five, it might be time to see whether your dreams are similar to those of other people or whether they are different in important ways. To do this, you will need a guide as to what other people's dreams are like. The very best guide I know of is a book by Calvin S. Hall and Robert L. Van de Castle entitled *The Content Analysis of Dreams.*

A few words about the book: the authors collected a large number of dreams from students in psychology classes at Western Reserve University and Baldwin Wallace College. The students were asked to describe their dreams as fully as they remembered them. In writing down their dream reports, the students were instructed to describe the setting of the dream, to mention the people who were in the dream, their own feelings during the dream, whether the dream was pleasant or unpleasant, and to tell exactly what happened during the dream.

In carrying out their analysis, Hall and Van de Castle chose a sample of 100 male and 100 female students who had provided a series of between 12 and 18 dream reports. They excluded any dream with fewer than 50 words or more than 300 words and then randomly selected five dreams from each of the students. This left them with a sample of 1,000 dreams to work with.

Much of their book explains in detail how they developed a system for analyzing the content of dreams. What does content analysis mean? In essence, it means looking at material that is qualitative in nature (for example, a diary, a series of letters, a book, a newspaper or magazine article) and making assessments of what you read. The judgment may be entirely qualitative (the book was about such and such and included photographs and references for further reading, etc.) or it may by quantitative (numbers which can be used for statistical analysis). An example of a qualitative judgment may be the judgments of an officer in the State Department Foreign Service stationed in Moscow or Beijing or Paris, who reads the daily newspapers and then writes down his assessment about what people are reading in these capital cities and what is being said in those articles. To be quantitative, the Foreign Service officer might tally the number of times the United States was written about in a favorable or unfavorable light.

Content analysis in the social and behavioral sciences is often quantitative and usually a very systematic approach, beginning with setting up categories for analysis and then looking through the material of interest for the presence or absence of what is described in the categories. The content analysis may be refined by the use of rating scales (e.g., the degree of favorability) to evaluate the material. The trick is to have meaningful categories to look for as you read and to make sure that these categories are defined clearly enough so that two people who read the material independently will make these judgments with a high degree of agreement.

Hall and Van de Castle did a very thorough job of developing categories for dream analysis, carefully defining what would be included in the category, and of providing examples for the analyst to consider in making his or her judgments, and, importantly, checking for agreement between independent analysts to see how reliable these assessments were. Because they did such a good job, researchers have used these categories to analyze dreams over the years.

The Hall-Van de Castle system to classify and score the content of dream reports included ratings for settings, characters, actions and interactions, emotions and many other things in the dreams. Consider settings, for example. The analysts began by deciding whether the dreams were indoors or outdoors. Characters were identified by whether they were acquainted with the dreamer and by age, gender and ethnicity. For emotions, the analysts had to decide whether there was anger, apprehension, happiness, sadness, etc., in the dreams.

These examples indicate just the beginning of the analyst's tasks, for there were many subcategories included for each classification, which permitted an increasingly detailed and finer analysis of what was reported in the dream.

While I have a very high regard for the work of Hall and Van de Castle, there are a few drawbacks in using their data for comparing your dreams with the dreams of others. Probably the most important is that *The Content Analysis of Dreams* is an old book published in 1966 and the dream reports that it draws on are even older. They were collected during the years 1947 to 1950. Because of its age, the book is not easily obtainable in bookstores or libraries, although I understand there is an Internet version available. The fact that the dream reports were collected well over half a century ago raises the caution that some of the statistics presented about the content of dreams may no longer be valid in the world we live in today. While research suggests that that there is considerable stability in the content of dreams over time, more than half a century is a long time, and there has been an enormous amount of change in the way people live. The demographics of the United States have shifted from a predominantly white population with a cultural heritage emanating from Europe to an international population with cultural traditions coming from all over the globe.

In addition, rapid development of technology following the widespread availability of computers, with ever-improving mobile devices, has changed the way we work, interact socially, entertain ourselves, and take in and process information from the world around us. The world of my childhood seems like a very different place from the world in which children are growing up today. Since dreams are largely based on autobiographical memory, the typicality of dreams of 1948 and those of today may differ. The research that we have available suggests that there is probably a general similarity between the dreams analyzed by Hall and Van de Castle and dreams more recently recorded, but I wouldn't dismiss the possibility of significant differences emerging. The children of 1948 couldn't dream about communicating with their friends on a smart phone because smart phones didn't exist. Many children in 1948 would be unlikely to include friends of different races in their dreams because, in much of the country, schools were segregated. The children of 1948 would be more likely to include a stay-at-home mom in their dreams than the children of today would, as contemporary mothers are much more likely to work outside the home.

Another drawback in relying on the statistics presented in the content analysis of dreams is that the sample was made up of college students. We know from research that the dreams of older people may be somewhat different from those of younger people, so a sample of college students, particularly in the 1940s, is unlikely to be representative of the nation as a whole. Moreover in the 1940s, a much smaller segment of people in their teens and early 20s went to college than today. In comparison to today's college students, the young people who went to college in the 1940s were more likely to be white, male, and better off economically. The community colleges that offer higher education today to so many young men and women hardly existed in the 1940s.

With these caveats in mind, let us now look at some of the statistics Hall and Van De Castle presented in their book. Let's first consider settings. When I think of a setting, I think of reading a novel or watching a movie. A gifted novelist paints a word picture of where a story happens. It might be a small town in pre-civil rights era Alabama, as in the Pulitzer Prize–winning novel *To Kill a Mockingbird* or the grim bastions of the land of Mordor in Tolkien's *Lord of the Rings* trilogy. The 19th-century English novelist of country life, Thomas Hardy, wrote page after page describing a heath in his novel *The Return of the Native.* Consider a movie which, of course, is more like a dream. As you watch a movie, you usually have a very good idea of where the story happens, even if it's in a galaxy far, far away as in the *Star Wars* franchise. When it comes to dreams, there is almost always a setting, although sometimes it may be unfamiliar. Can you imagine a dream without a setting? That would be surrealistic. Dreams without a setting are rarities. Of the 1,000 dreams analyzed by Hall and Van De Castle, only 19 were coded as having no setting.

The 1,000 dreams in the sample contained almost 1,300 settings, so it is clear that some dreams included more than one setting. While some shift in the scene occurred in some dreams, the average number of scenes for both men and women was about 1.3, which indicates that most dreams did not rapidly move from one scene to another, so if you frequently have dreams which rapidly shift in settings, that would be a benchmark of atypical dreaming, as dreams usually take place in only one or two places.

Are dreams more likely to occur indoors or outdoors? For men, the data suggest that both indoor and outdoor settings seem about equally likely. For women, dreams are more likely to take place indoors. Perhaps these gender differences simply reflect life in the late 1940s, when the role that most women assumed in our society was taking charge of the home and rearing children.

At times, our dreams occur in settings that seem totally unfamiliar. A friend recently told me about a dream he had in which he was in a very large house that he had never been in or even seen. Both the house and the adjacent grounds were totally unfamiliar to him. For some people, the setting of a dream may be distorted or even bizarre. We might wonder whether the settings for our dreams are more likely to be a familiar place or in some place that is unfamiliar. Looking at the data, for both men and women the answer is dreams tend to occur in a familiar place rather than in an unfamiliar one. For men the ratio of the familiar to the unfamiliar scene was 1.61. For women the ratio was higher, 3.71.

Once again, these different ratios seem to reflect the kind of cultural exposures that boys and girls had in the 1940s when these dreams were collected. The students in the sample were children during the Second World War, a time when men by the millions were going off to war into a challenging and potentially deadly environment while women for the most part remained at home, taking care of home and family, or working in war-related industries. The books that boys and girls read were very different in the 1930s and 1940s. The books boys read were more adventurous. The games boys played were far more physical and competitive, and in general, the role expectations for one's future life in those years were very much gender determined.

Let's turn to the characters in your dreams. Since dreams tend to utilize autobiographical memories, we would expect them to include frequently those individuals who are most important to you. On occasion, we might expect your dreams to include your spouse and children if you have them, or people who are in or have been in an important relationship in your life, such as your mother and father, your siblings if you have them, possibly an important relative like a grandparent, and, of course, your friends. We might expect these people to be supporting actors in your dreams, with you yourself in the starring role. In addition to familiar people, our dreams will at times include people who are complete strangers to us.

Now let us look at some statistics provided by Hall and Van de Castle to see how this all shakes out. First, can you imagine a dream without any characters besides the dreamer? It would be an unusual sort of dream but it does happen, although infrequently, about five percent of the time. Typically, there are two or three characters in the dream, including the dreamer, and this is another benchmark to remember. If you usually have lots of characters in your dreams, that is atypical.

Occasionally, people do report dreams with a whole host of characters. Can you imagine a dream with 10 characters in it? That would be something like creating your own nightly television soap opera. I picked the number *10* because that was the highest number noted in the sample of 1,000 dreams.

Women tended to have more characters in their dreams than men. As we have mentioned previously, men were more likely to dream about men than women (587 to 286) while women were more even in the gender of their dream characters (507 women, 547 men). I wonder whether this would still be true in the time we live in today.

Let's talk in more detail now about the characters in dreams. To get in the right frame of mind, think of a student in his or her late teens or early twenties and imagine who would be the important people in his or her life. Then imagine who would be the players in their dreams. If you suspected that the players would be their friends, you are probably right. In their analysis of dream characters, Hall and Van de Castle used a broad category of adults who were not family members or relatives, or professionals like doctors or dentists or teachers, just people who they knew who were *not* children. I suspect for students attending college, most of the people who would fall into this category would very likely be fellow students or friends back home. Well over 800 of these characters appeared in the dreams. Therefore, I think it is likely that these college students dreamed mostly about people who were friends and acquaintances.

How about family members? How often do they appear in the students' dreams? Hall and Van de Castle's data indicate that there were about 300 family members mentioned in the sample of 1,000 dreams. The family members most often reported in the dreams were the dreamer's parents. Fathers and mothers were mentioned about equally often by the men while the women in the sample mentioned their mothers more frequently than their fathers. Brothers and sisters were mentioned less often than parents and relatives still less often.

The presence of strangers in dreams, people the dreamer did

not recognize, strikes me as intriguing. There were nearly 500 of these unknown people in the sample of 1,000 dreams. That's a lot more than family members. Who were these strangers? What were they doing? Why were they there? While Hall and Van de Castle's data tell us that most of these strangers were adults, the data does not shed much more light on them. I regret that. It might have been very revealing. There was once a popular song called "Strangers in the Night" and I'm afraid our knowledge of their role in our dreams does not go much beyond the title of the song.

Let's now turn to social interactions. Hall and Van de Castle were primarily interested in three types of social interactions that occurred in dreams. These were aggressive, friendly, and sexual. You might wish to stop here and ask why only three categories. Surely not all encounters with other people fit neatly into these three categories. Consider, for example, a mother holding and cuddling her newborn baby. Certainly that's friendly behavior, but it's a lot more than this. It's nurturing behavior combining love and tenderness. Or consider the reaction of the students in a classroom when the teacher asks them to write a composition. Complying with the teacher's instruction does not fall neatly into any of Hall and Van de Castle's categories. The students comply with the teacher's instructions because they feel they have no real choice to do otherwise. Life's interactions are far too subtle to put into three boxes, and this includes the interactions that occur in dreams.

Having said that, aggressive behavior is certainly an important aspect of interactions both in real life and in dreams. We know that the amygdala is active during dreaming and that it is an important control center for anger, so we should not be surprised to see fairly significant numbers of dreams containing aggression, and this is what Hall and Van de Castle reported. Among the sample of 500 dreams collected from the male subjects, aggression occurred in nearly half of them, and the number reported for females was almost identical. There was more than one instance of aggression in many of the dreams. In most cases, the dreamer is not the aggressor. The dreamer is more likely to be the victim. There is also a smaller number of dreams in which the dreamer merely witnesses aggression taking place in the dream.

If you are usually the victim when there is an aggressive act or statement in your dream or merely a bystander looking on, that would not be atypical. To be the frequent perpetrator of such acts in your dreams might be something to think about.

Not surprisingly, for both men and women, aggression in dreams is more likely to come from a man than from a woman, and it is interesting that this aggression often comes from people who are unfamiliar to you. Once again, strangers had an important role in our dreams.

Do you recall dreams in which you were with people who acted in a friendly manner? If so, these dreams are probably atypical. They don't happen that often. Hall and Van de Castle tabulated the number of times friendliness was reported in each of 1,000 dreams and found that the most frequent number they came up with was *zero*. Most dreams did not include friendly acts, and those that did typically had only one friendly act in the dream.

When there were friendly acts in the dreams, were they instances of give and take, of mutual sharing? Were they reciprocal acts? The answer is *rarely*. In the sample of 1,000 dreams, only 10 were identified as evidencing reciprocal friendliness.

If your dreams frequently include acts of shared friendly behavior, that certainly appears atypical, but if they do, consider yourself lucky.

For the young men in the sample, acts of friendliness, when they did occur, were offered by characters in the dream who were unfamiliar to the dreamer about as often as by people the dreamer knew. This was not true for the women in the sample. The dream characters who were involved in friendly acts were twice as likely to be familiar as unfamiliar. These results seem to tie in with the overall picture one sees in Hall and Van de Castle's data that the dreams of young women in this mid–20th century sample tended to draw on the familiar in life.

Returning to the frequency of sexual activity in dreams, when Hall and Van de Castle made tabulations for the number of dreams in which any aspect of sex occurred—whether it was sexual intercourse, kissing, any form of fondling or even verbal sexual overtures—the figures for men, 58 out of 500 (12 percent), were higher than for women, 18 out of 500 (4 percent). However, I think it is important to note that Hall and Van de Castle themselves doubted the accuracy of these figures, suggesting that neither the men nor the women were completely candid in reporting sexual activities in their dreams and that the women were probably less candid than the men. I share their skepticism. In the late 1940s, American society took a far less permissive attitude toward sexuality than today. It was difficult to obtain a copy of a novel in which sex was explicit and there was a strict code in Hollywood against showing anything that was more than suggestive. And you simply didn't see sex on television. One should take these figures with the proverbial grain of salt.

If there were any aspects of sex in the dreams that were reported, even verbal sexual overtures, we may ask who was involved. We are dealing here with very small numbers but it is nonetheless interesting that for the men in the sample there was a slight tendency for the action to involve someone who was unfamiliar. For the very few women who chose to report any aspect of sexual contact, the person involved was familiar.

Considering the more mundane aspects of dreams, the authors presented an alphabetical list of 1,170 objects mentioned in the dreams and noted the frequencies with which they appeared. Among the most often reported objects were geographical sites such as cities, avenues and streets, buildings such as houses and the rooms within the buildings, as well as stairs and windows and furniture such as tables. Not surprisingly, automobiles were mentioned frequently in the dreams. Parts of the human body like heads and hands were occasionally mentioned, but less often than the above, and there was almost no mention of the areas of the body associated with explicit sexuality.

If you often have dreams of being out of doors, you will be surprised to learn that there was hardly any mention of grass and only a few mentions of birds. Only trees made it into double digits. Think about roses for a moment. Remember the oft-quoted line about taking time to stop and smell the roses? While the admonition suggests that we fail to do this often enough, it seems we do not dream about them at all, or, at least, hardly ever. Only two instances of roses were coded in 1,000 dreams.

Let's consider activities in the dreams. Think about your own dreams for a moment. Do your dreams frequently include conversations? Or, more typically, are your dreams ones in which you are silent, perhaps taking some action or simply looking at what is happening around you? For the men in the sample there was a slight tendency for the actions that occur in the dreams to be physical rather than verbal. For the women in the sample, we see the opposite pattern with more conversation than physical activities. Once again, this seems consistent with the view that dreams in large part reflect the realities of daytime experience.

There is some interesting data on the number of activities in a dream. Typically, for both men and women, there was more than a single activity in a dream. Single activity dreams were rare. Dreams in which there were three or four or even six or seven activities are quite common. However, as noted previously, these activities tend to stay in one place in the dream. Only about one in ten of the activities reported in the dreams of both men and women involved a change in location. The wildly changing scenes of shifting locations was not a common occurrence.

When Hall and Van de Castle studied emotions in dreams, they did not single out the emotions aroused in nightmares and night terrors. Nightmares and night terrors do not even appear in their book's index. Yet, the emotions raised in these dream experiences are by far the strongest emotions experienced in dreams. Instead of considering these emotions separately, Hall and Van de Castle subsumed them under the category they called "apprehension." The emotions listed in this category ranged from "terrified" and "horrified" to the more mild term "concerned." Apprehension also included "remorseful," "sorry," and "ashamed." We all know that there is a marked difference between feeling terrified during a nightmare and feeling concerned and both terms are certainly different from feeling ashamed, which seems to me like it should have been included under "guilt." We have to keep this in mind when we look at the statistics relating to apprehension, for we are likely dealing with a mixed bag, and the statistics offered may well be inflated.

In addition to apprehension, the dreams were scored for "anger" using such words as "annoyed," "furious," and "mad." The dreams were also scored for categories of "happiness" and "sadness." For reasons

I have difficulty understanding, the authors included a category called "confusion" along with these other, more clear-cut categories of emotion. With these reservations, let us see what the researchers found and ask ourselves once again if these levels of emotions are similar to what you experience in your own dreams.

First, many of the dreams reported by the college students did not contain any clear-cut emotions. This was particularly noticeable for the men in the sample where the total number of emotions noted in their 500 dreams was only 282. The most frequent of these emotions noted was apprehension followed by confusion, happiness, anger and sadness. For the women in the sample, there was a total of 420 incidents of emotion scored with apprehension by far the most frequent. Apprehension was followed by happiness, confusion and a virtual tie between sadness and anger.

When considering the typicality of your own dreams, you might ask yourself whether your dreams frequently contain emotions and whether apprehension is the most frequent emotion that you experience. While a *yes* to both questions is not as definitive for men as it is for women, for women, affirmative responses suggest that as far as emotions are concerned, your dreams are not that different from the dreams reported by these college students.

Have you ever had a dream in which you were attempting to solve a problem? It might have been a problem relating to homework you were given at school or a problem you had been working on at your job. It may even have been an interpersonal problem that had been causing some concern, perhaps one in which you were trying to improve an important relationship that was becoming difficult. If you had such dreams, what was the outcome typically like? Did you experience success, or were the efforts in your dreams unproductive, unsuccessful? When Hall and Van de Castle coded the sample of 1,000 dreams, they found that the number of dreams in which success was reported and the number of dreams in which failure was reported were not that different. There were 113 of the dreams coded as including success and 126 as including failure. The experience of both success and failure were slightly higher in males. In almost all cases, it was the dreamer who was experiencing success or failure, not other characters in the dream.

One final note: the dreams were coded as to whether they included good fortune. Only 58 of the 1,000 dreams were coded this way, so it looks once again that pleasant dreams are far from the rule. The lyrics in the beautiful song in the classic Broadway hit *The Music Man* seem more of a wish for most people than a reality. For the sample of students studied by Hall and Van de Castle, dreams were rarely sweet, and if we venture to the other extreme, for most people, dreams are not typically nightmares, either. Dreams are a mixed bag, something in between, with some level of apprehension not uncommon.

This mixed bag of our dreams is likely to include dreams that occur during REM sleep and NREM sleep. The dreams that occur shortly before you wake in the morning are likely to be REM dreams, but it is unlikely that when you think about your dreams during the following day, you will be able to distinguish with any accuracy which of your dreams occurred during the different kinds of sleep. Certainly over the long term, your recall of your dreams will include both the more mundane NREM dreams as well as the more interesting dreams of REM sleep.

If most of your dreams are like those of the people in the Hall and Van de Castle study, you will usually have dreams that include a number of actions, but you are unlikely to experience frequent changes in settings. There will usually be only a small number of characters in your dreams, probably friends and family members, perhaps parents. If you are female, dreams are most likely to occur indoors and in familiar places. Interactions with other characters may often involve aggression, and when hostile acts and speech occurs, it will most likely be against yourself. Acts of kindness and friendliness are not likely to be frequent and the emotions you experience in dreams are more likely to be uncomfortable than comfortable. The typical dreams that most people experience are unlikely to be pleasant experiences.

You may well ask, what happens when my dreams fall outside these statistical parameters? What if my dreams are frequently atypical? Am I likely to remember these dreams better? Do such dreams have special meaning or significance? Based on the research that I have seen, I am not able to offer a categorical answer to these questions, although studies suggest that if the unusual elements of your dreams are *bizarre*, your chances of recalling the dream are increased. Whether unusual dreams have special significance for the dreamer is an important question, probably a very deep one. Intuitively, I suspect that is the case, but I would like to see more direct studies of this question before I conclude that it is so. As we move ahead and consider the different methods of dream interpretation, I think it might be a good idea to keep such unusual dreams in mind. Perhaps the possible significance of these dreams may become more apparent as we delve deeper into methods of dream analysis.

12

The Approach to Dream Analysis of Sigmund Freud

Let us begin our exploration of contemporary methods of dream analysis with the psychoanalytic approach of Sigmund Freud. Freud's method of dream analysis drew heavily on a procedure he developed called *free association*. Freud developed the method of free association while he was working with patients with psychiatric symptoms such as hysterical paralysis and obsessional ideas. He asked the patients to tell him every idea or thought that occurred to them in connection with some particular subject they brought up during the therapy session. He found that the material that came forth during free association was productive, helping him go back through the patient's mental life to the issues where the symptoms had their origins.

There came a time when Freud decided to try to apply the method of free association to his patients' dream reports. In doing so, he had to instruct his patients that when they considered what to tell him, they should pay serious attention to all of their own ideas and thoughts and to eliminate any screening process they normally used in everyday conversation. They were told to avoid sifting through their thoughts before uttering them. They were to report whatever came into their minds and not to suppress anything simply because it seemed unimportant or irrelevant.

In using free association in interpreting dreams, Freud felt it was important to break up the dreams into their separate parts. Freud found that if he asked patients to free associate to an entire dream, he did not get very much that interested him. However, when he put the separate parts of the dreams before his patients, one at a time, he got much more useful material.

Essentially, this procedure asked patients to free associate to the different sections of the dream, verbalizing anything that came to mind, and it became the analyst's task to put this material into a coherent picture using Freud's theory of psychoanalysis as the bedrock for interpretation.

Let's see how Freud analyzed dreams using a case study. When we look at the index of Freud's book *The Interpretation of Dreams*, we find a list of dreams that we can choose from. The first list is of Freud's own dreams, the second of the dreams of other people. The titles given to some dreams are interesting, even intriguing. Among the titles of dreams of others were "Black luster dress," "Children grew wings," "Diving into a lake," "The language of flowers," "Barrister's lost cases," and "Dead bodies being burnt." The latter two sound like cases from a detective novel, and when one thinks about it, the way Freud went about analyzing dreams reminds one of the methods used by classic fictional detectives such as Sherlock Holmes, who had nothing like the crime labs of today to work with, but like Freud, had to rely almost totally on their brains to piece together bits of information and come to conclusions. Some of the titles given to Freud's own dreams are "Riding on a horse," "Botanical monograph," and "Reconciliation with a friend." In selecting a dream to illustrate Freud's method, I chose the dream "Irma's injection," a dream Freud himself selected to demonstrate his method of dream interpretation. Because Freud's presentation of the dream and its interpretation is quite lengthy, I have taken the liberty of condensing what he wrote, but I trust I have still fairly illustrated the way he analyzed the dream.

Freud had been treating a young woman, a family friend named Irma, who had been suffering from physical symptoms that Freud had diagnosed as hysteria, indicating that he believed the causes of her symptoms were psychological. Freud had been treating her with his newly developed technique, psychoanalysis. However, his treatment of Irma had only been partially successful. When one of Freud's colleagues, a man named Otto who also knew the patient well, told Freud that Irma was better but not quite well, Freud was annoyed that his colleague had given his treatment a less than stellar evaluation.

That night Freud had a dream in which he and Irma were in the same room, a hall in his house where he was greeting people—a gathering of friends. He said to Irma, "If you still get pains, it's really only your fault." Irma replied that she was experiencing great pains in her throat and stomach that were choking her. Alarmed, Freud thought that he must have missed some physical disease and began to examine her. As he did so, he saw that she looked pale and puffy and there was a big white patch in her mouth as well as whitish gray scabs. Freud quickly called in another doctor, a man referred to as Dr. M., to take a look, and Dr. M. said that there was no doubt it was an infection. Interestingly, in the dream, Dr. M. looked very different from the way he usually did: he was clean-shaven, very pale and walked with a limp. Wondering what might have caused the infection, Freud thought that it was Otto's fault because he had given her an injection and that it had been done thoughtlessly and probably the syringe had not been cleaned.

In analyzing the dream, Freud broke it up into segments. He began with the hall in the house where he was staying and the guests who were received there. He remembered that on the previous day his wife had told him that she would be expecting a number of friends to celebrate her birthday, including Irma. He concluded that his dream was anticipating this occasion.

He then thought about his criticism of Irma, that if she still had pains it was her own fault. He remembered that he had held the view at that time that his responsibilities in treatment were fulfilled, when he had informed the patient of the real (psychological) causes of her symptoms and that he was not responsible for whether the patient accepted his interpretation or not. He had since disregarded this idea.

In thinking about his alarm that he might have missed a physical illness, Freud thought this concern was a perpetual source of anxiety to a specialist whose practice was almost limited to patients with psychological problems. It further occurred to him that he might be wishing that he had indeed made a wrong diagnosis, for if he had, he could not be blamed for not curing her.

When he examined her throat, he was reminded of two other women he thought were hysterics; one like Irma suffered from hysterical choking. In considering these women, he suspected that maybe he would have liked to exchange Irma for one of the others who was more intelligent and might respond better to his treatment.

The white patch in her throat reminded Freud of a serious illness his own daughter had experienced about two years earlier, and the scabs reminded him about the worry he had experienced about his own state of health. He had been using cocaine—which at the time was a drug that we knew little about—to reduce some nasal swelling and he had heard that one of his female patients who followed his example had developed serious problems. This use of the drug brought serious reproaches from members of the medical community to Freud.

My comment: you can see from this example that a small fragment of the dream could produce a trail of associations, one thought leading to the next. A small fragment is a far cry from a full-fledged event and a trail of associations can lead further and further away from the actual content of the dream.

His calling in another doctor, Dr. M., to examine Irma brought to mind a tragic event in his own practice where he had produced a severe toxic state in a female patient by repeatedly prescribing what was at the time regarded as a harmless remedy. Interestingly, the patient who succumbed to the drug had the same name as his eldest daughter.

In thinking about Dr. M., who looked pale, clean-shaven and walked with a limp, Freud thought of his older brother who looked like the Dr. M of the dream and had recently walked with a limp caused by arthritis in his hip. Freud then recalled that he had a similar reason for being annoyed with each of them, as they had both rejected a suggestion he had recently made to them.

His thoughts about the part of the dream where Otto had given an injection to Irma reminded Freud that Otto had told him that during his stay with Irma's family, he had been called into a neighborhood hotel where he had in fact given an injection to a person who had fallen suddenly ill. This injection reminded him of a friend who had poisoned himself with cocaine, disregarding Freud's advice to take the drug orally, but had instead injected the drug.

While there is a great deal more to the analysis of the dream, one can see already that Freud's trains of associations had uncovered hostility on his own part as well as a need to justify his professional conduct. In regard to the latter, Freud noted that the dream absolved him of responsibility for the persistence of Irma's pains by placing the blame on Otto. Freud observed that the dream had not only placed the blame on Otto but gone beyond this straightforward means of acquitting Freud, producing a whole series of reasons why Irma's pain was not Freud's fault. Freud believed that the dream presented a state of affairs as he would have wished it to be. He concluded that the content of the dream was a fulfillment of a wish and the motive for the dream was a wish.

The analysis of the Irma dream was a milestone in Freud's development of his method of dream interpretation. In his seminal work *The Interpretation of Dreams*, Freud cites the analysis of this dream many times. Over the years, his followers have written papers about it, studying what Freud wrote and commenting on it in some ways that resemble a theological scholar interpreting a verse from the Bible.

Freud's analysis of Irma's dream calls attention to the way he considered the details—the fragments of a dream important in his analysis. Recall the excerpts such as the white patch on Irma's throat and Dr. M.'s having shaved. Other practitioners of dream analysis might have considered such details trivial and ignored them, but not Freud.

This case also demonstrates the way Freud free associated to these dream fragments, giving free rein to whatever memories and thoughts that arose. Turning the mind loose, permitting it to roam freely over the fragments of the dream, was a critical part of Freud's method of dream analysis. Making sense of these associations and tying them together was a task for the analyst, presumably relying heavily on Freud's theories of dreams.

When Freud developed his theory of dreams and his method of free association as a way to unravel the meaning of dreams, he was working from a psychological perspective, a model of the mind as he envisioned it, rather than working from a perspective of how the brain works to produce dreams. He did not have available anything like the knowledge we have today about the activities of the brain. The technology was simply not there. The chief agent in Freud's attempt to describe and understand the actions in dreams was a hypothetical construct he called the *dreamwork*. Freud conceived of the dreamwork as an agent that could turn the events of real experience into what we experience as dreams using such mechanisms as condensation, displacement, and symbols.

While the product of Freud's dreamwork might possibly be construed today as manifestations of some of the activations and deactivations of areas of the brain discussed earlier, there is little we have seen to date in the workings of the brain that confirms the relentless war going on between unacceptable wishes and censoring activities, which is the heart of Feud's theory of dreams. Moreover, the actions of the brain seem secondary in Freud's thinking that dreams were in essence psychologically determined. Freud believed that the content of our dreams is determined by recent events that trigger and then combine with wishes and conflicts hidden in the mind. This is, of course, a far cry from the activation-synthesis hypothesis that looks at dreams as basically determined by actions of the brain.

A contemporary writer described the Freudian point of view this way: "From the psychoanalytical point of view, the dream transcends neurobiological knowledge, and looks like a process of internal activation that is only apparently chaotic, but is actually rich in meanings, arising from the person's affective and emotional history." Put more simply, dreams have meanings that tell us important things about the person's psychological makeup, and if we are smart enough, we can figure out what these meanings are.

Freud's idea to use free association to unravel the meaning of dreams has always seemed to me a giant leap forward when compared to the arbitrary assignment of meaning to objects and actions we see in dream dictionaries. Free association usually leads to autobiographical memories that can be examined by a therapist or by the dreamer to assess their importance in his or her life. There is obviously a difference between unrestricted free association, which Freud espoused—letting one's mind wander from one idea to another, taking one further and further away from the actual content of the dream—to a more controlled association process in which one looks at particular events in one's life that are associated with the dream. We shall examine modifications to Freud's free association method that have been developed by others as we proceed.

Are the conclusions that Freud and his followers reached by

using free association to dreams and then interpreting them valid? Are these interpretations valid insights about the issues troubling the dreamer? Freudian-oriented therapists seem to think so because they see benefits, even cures, in their patients after psychoanalytic treatment. However, as we have indicated before, people get better after they engage in almost any form of psychotherapy—whether it is psychoanalysis, behavior therapy, cognitive behavior therapy, interpersonal therapy, or other varieties of therapy. This improvement, in almost all forms of psychotherapy, reflects what happens in the developing relationship between the therapist and patient. When patients have an opportunity to talk about what is really troubling them to someone who patiently listens, is understanding, and can offer what is accepted as thoughtful feedback, this can be therapeutic in itself, regardless of the theoretical orientation of the therapist.

It is difficult to find examples of serious objective research to demonstrate the validity of Freud's method of free association to dreams. For many years, I have had the impression that the psychoanalytic community, persuaded by Freudian theories, or at least comfortable with them, has shown little interest in subjecting Freudian ideas to rigorous testing, certainly not research that meets traditional scientific standards. Realizing that I might be biased in this regard, I ran a quick check by entering the term "psychoanalytic study of dreams" into the National Library of Medicine's database (pubmed. gov) and looked through several pages of titles of articles. Some of the titles looked very interesting, although on occasion, I came across a title that made me shake my head. There were indeed a number of titles discussing Freud's original researches. All in all, however, I found few titles that indicated research meeting the usual scientific standards for objectivity, replicability, the use of control groups, and the application of statistical data to test for the effects of chance.

If this is indeed the standard of proof followed by the adherents of Freudian psychology, we really shouldn't expect much in the way of hard scientific data from the psychoanalytic community in testing Freud's ideas, and, indeed, there isn't a great deal. Certainly, the task of investigating material presumed to be unconscious is a difficult one, but there are examples, including my own work on dream symbolism, which show that objective research is possible, and I would like to conclude by citing an interesting study dealing with free association to dreams carried out in the 1990s by researchers in Germany.

These researchers carried out an experiment to examine Freud's premise that the elements of the dream can be a kind of gateway to the unacceptable wishes and counter forces in the unconscious and that the process of free association to the dream elements would encounter increasing resistance as the use of free association proceeds. The researchers asked a group of female subjects to associate to parts of their own dreams and to those of another person, a control subject. While making these associations, the subjects were asked to press a button when they experienced feelings of uneasiness during the free association task. The researchers took measures of skin conductance reactions during the procedure. Skin conductance is believed to be a measure of psychophysiological activation. The researchers reported that not only were these measures of activation greater when the women made associations to their own dreams than those of the control subject, but their subjective feelings of uneasiness were more pronounced as well when they were associating to their own dreams. These findings seem consistent with Freud's views that resistance will arise in dream interpretation and especially as significant and uncomfortable ideas are unearthed.

This seems to me a good experiment and I only wish there were many more like it.

13

Dream Analysis According to Carl Gustav Jung

Carl Jung wrote a great deal on dreams, most of it in the first half of the 20th century. We are fortunate that a lot of Jung's ideas about dreams and dream analysis have been collected in a single volume entitled *Dreams*, published by Princeton University Press in 1974. I found the book easily available in my public library and it served as the major source for my description of Jung's views of dreams.

I might also mention another source I found useful in presenting an overview of Jung's thinking on dreams, an article by Caifang Zhu entitled "Jung on the Nature and Interpretation of Dreams: A Developmental Delineation with Cognitive Neuroscientific Responses." Zhu reminds us that Jung went through several phases in his thinking about dreams, first championing Freud's ideas about dreams, then developing his own theories, particularly his theory of compensation, and then becoming more interested in symbols and in Eastern philosophies. Like many others, I find his theory of compensation in dreams his most interesting contribution, and I will focus most of my attention on it as I look into his approach to dream analysis.

Like Freud, Jung believed in the importance of the unconscious, and like Freud, Jung's idea about the unconscious was a hypothetical construct, something he fashioned in his thinking while considering such seemingly disparate activities as dreams, mythology and the use of symbolism, both in earlier times and during the present. While disagreeing profoundly with Freud about the nature of the unconscious, he insisted upon its importance and agreed with Freud that dreams offer an important avenue to delve into the unconscious. He wrote, "Nobody doubts the importance of conscious experience; why then should we doubt the significance of unconscious happenings?"

Before delving more deeply into Jung's approach to dream analysis, let's recall his ideas about compensation, for they are a key to understanding where Jung will be taking us. While Jung's ideas about compensation were nuanced and shifted somewhat over time, he tended to view the entire human psyche as a "self-regulating system that maintains its equilibrium just as the body does." He believed that every process occurring both in body and mind could go too far in one direction and that inevitably led to a reaction he called "compensation." This notion immediately brings to mind Newton's third law of motion. In Jung's thinking, this compensatory reaction not only held true for such well understood regulated activities of the body such as normal metabolism, but it also was a basic law of psychic behavior. He believed that too little movement in one direction tends to bring forth too much in the other, and importantly, this was true for the relationship between conscious and unconscious activities. He argued that when we begin to interpret a dream, we should ask what conscious attitude it compensates for. He believed that every dream is a source of information that can tell us useful things that we can pay attention to, and by doing so, may help keep ourselves on an even keel.

In applying the idea of compensation to dreams, he argued, "no amount of skepticism and criticism has yet enabled me to regard dreams as negligible occurrences. Often enough they appear senseless, but it is obviously we who lack the sense and ingenuity to read the enigmatic message from the nocturnal realm of the psyche." He stated that consciousness acts upon our nightly life just as much as the unconscious affects our daily life. In other words, one influences the other in a compensatory way.

Let's think about this for a minute. Perhaps you are a student in high school or college and you have been studying very hard for exams or you have been working at a job that has kept you very, very busy, putting in long hours during the week and sometimes even on the weekend. You might be feeling tired, a little worn, even irritable. At nighttime, you have had dreams about friends. In one dream, you're sharing a meal with them at a nice restaurant, enjoying a drink while you are engaged in a pleasant conversation punctuated with smiles, laughter, and good feelings. When you think about these dreams, you recall some good times spent with these friends. Jung would say that this dream is acting in a compensatory way, recalling the good feelings when your life was more in balance. Now, let us imagine how you feel the next morning. Maybe your mood has improved. Maybe you have thought about calling up some of your friends and making plans to spend some time together. While you are thinking about this, perhaps the idea has crossed your mind that you need to break up your work routine with a little diversion and add some more enjoyment to your life. If these thoughts give wings to behavior, to doing these things, you could say that the pendulum in your daily life has swung in the other direction. Your thoughts in your waking life about good times and your renewed emphasis on interaction with your friends may now be compensating for the excesses that were characterizing your working life. As Jung would put it, your psychic life is now more in equilibrium. Of course, there is the possibility the pendulum could swing too far and you might begin to dream about not getting a good grade at school or a promotion at work.

In his article, Zhu cited an instance where Jung applied his theory of compensation to one of his own dreams. In the way of background, Jung was treating a patient described as a highly intelligent woman. During therapy, Jung's analysis of her dream proceeded well at first, but in time he got stuck with his interpretation and noticed some shallowness in his conversation with the patient. He decided to discuss this with the patient. On the night before he was to meet with her again, he had a dream in which he was walking down a highway through a valley and came upon a steep hill. At the hill's summit, he saw a castle, and on the highest tower on the castle, he spied a woman sitting on what looked like a balustrade. In order to see her better he had to bend his head far back, and indeed, he awakened with a crick in the back of his neck. He said that even in the dream he recognized the woman as his patient.

In interpreting his dream, Jung thought that if he had to look up at the woman, his patient in the dream, it was likely that in his waking life, he had probably been looking down on her, not only intellectually but also morally. He interpreted the dream as a compensation for what his attitude had been toward her in his waking hours. Jung discussed the dream and his interpretation of it with the patient and immediately afterward there was a positive change in the course of therapy.

As stated previously, both Freud and Jung argued for the importance of dreams and for understanding the unconscious activities of our minds. However, they disagreed sharply about the nature of the unconscious and this difference influenced what they looked for in dreams and how they proceeded to analyze dreams. In an oblique reference to Freud, who viewed the unconscious as containing unacceptable ideas and wishes held down by censoring forces, Jung stated that he did not view the unconscious as a "demoniacal monster," but as a "natural entity" that was morally, intellectually, and aesthetically neutral. It was not by itself a negative thing—if one could refer to it as a thing—and it wasn't likely to be a source of problems unless what it contained was ignored or repressed. Jung's idea of the unconscious was a complex entity, and we shall have more to say about this when we discuss his views about symbolism and particularly the use of symbols in dreams.

In trying to analyze dreams, Jung disparaged the idea of starting out with a theory about the meaning of dreams and then trying to use it as a vehicle to interpret dreams. This could be taken as another knock against Freud, who had painstakingly developed an elaborate, comprehensive theory of dreams that guided his practice of dream analysis. For Jung, the first task was not to try to understand and interpret the dream, but, as he put it, "to establish the context of the dream" with meticulous care. Jung rejected Freud's approach to free association, which sets one upon a path emanating from every image of the dream, whether significant or obscure. Instead, he argued for a very careful look at particular images in the dream, directing his patients to bring up associations connected with these images. He dismissed Freud's reliance on free associations saying that such a procedure wouldn't help him decipher a dream any more than it would help him understand an ancient Hittite inscription. Jung's admonition was to stick as closely as one could to the dream images and probe deeply into them. When considering an object that his patient mentioned in the dream, Jung said that he would not be satisfied until he had learned as much as he felt he could about the object. Jung felt little hesitation in asking his patients to describe the dream object in more detail so he could more fully understand what sort of thing it was. In this way, he sought to establish the context of the dream image in its entirety.

Probing aspects of the dream was not out of line for Jung. On the contrary, it could be essential. The stereotypical view of a Freudian psychoanalyst listening, listening, endlessly listening, was not what Jung had in mind when it came to dream analysis.

So, as his patients offered their associations to their dream images, Jung might ask them questions about these associations in order to clarify them. Here is a brief example from one of Jung's illustrations of a dream analysis: a man told Jung that he had had a dream about the then-pope, Pius X, whose features were much more handsome in the dream than they were in reality. When the man mentioned that the Pope's nose was well formed and slightly pointed, Jung asked, "Who has a nose like that? Was there anything else noteworthy about the Pope's face?"

Jung made it clear that only after the associations to these dream images were examined in detail was it time to try to interpret the dream. As we discussed, Jung rejected the idea of beginning a dream interpretation with a theory. He looked at every interpretation as a hypothesis, something like reading an unknown text, and importantly, Jung believed that it was necessary to consider a series of dreams before one could place real confidence in the conclusions one could make about an individual. A single dream taken in isolation could not be interpreted with any degree of certainty.

When we inquire about how well Jung's view of dreams and dream analysis dovetails with the science about dreams presented in the first part of this book, we must remember that like Freud, Jung was not particularly interested in how the brain gave rise to dreams, for when Jung developed his major ideas, there was no way at the time to seriously study this question. How Jung's complex model of the unconscious, including the somewhat nebulous and certainly unwieldy concept of the collective unconscious would fit in with our current knowledge of brain activity during REM and NREM dreaming would be a very challenging endeavor and I think would be highly speculative at this point in time. So, I will leave this task to Jungian scholars and interested neuroscientists and plunge directly into the area of Jung's thinking that I find most relevant to the interpretation of dreams: his theory of compensation. What does available research suggest about this theory? What kind of additional research might be possible?

As I did with Freud, I began my search for empirical research designed to support Jung's theory of compensation by turning to the National Library of Medicine's online reference, pubmed.gov. First, I looked through all of the citations that emerged when I put in the search term "Jung's theory of dreams." There were three pages of citations, but only two references looked like traditional scientific studies. These were entitled "Compensatory Aspects of Dreams: An Empirical Test of Jung's Theory" and "Archetypal Dreams and 'Everyday' Dreams: A Statistical Investigation into Jung's Theory of the Collective Unconscious." Interestingly, both of these studies were carried out into the 1970s, which suggests that followers of Jung have shown little interest in testing his theories in an objective, replicable way, using the standards of scientific research. Much more likely to appear in the list of references were papers such as "C.G. Jung's Dream of Siegfried: A Psychobiographical Study." What I found for the most part were scholarly studies like the above written by adherents of Jung's theories, applying Jung's thinking to different aspects of our culture. That's all well and good, but not what I was looking for, research evidence supporting his theory of compensation of dreams. This conclusion was very similar to the one I advanced about Freud.

I made a modest check of this conclusion by looking at recent issues of the *Journal of Analytical Psychology*, which specializes in topics that appeal to Jungian scholars. The titles of the articles I saw, such as "The Realization of Absolute Beauty: An Interpretation of the Fairy Tale Snow White," would interest many, but would be of little value in proving or disproving Jung's theory about compensation in dreams. So, once again, I was left with the feeling that if one were to objectively assess the validity of Jung's compensation theory of dreams, the research would probably have to be done by someone other than the followers of Jung. Incidentally, the one study that sought to test Jung's theory of compensation in dreams concluded that the findings did not support the theory. The researchers concluded that dreams were not compensatory and did not differ substantially from conscious functioning.

While my direct search turned up little, it still seems possible that one could find evidence bearing on Jung's theory of compensation by meticulously looking through the findings of the hundreds of studies that have been carried out on dreams—even though the studies were not designed specifically to test the theory of compensation. A case in point is my own study of Freud's theory of dream symbolism. While not even thinking about Jung, we collected data that bears on his theory of compensation.

Let's backtrack for a moment. Recall that Jung postulated that the content of dreams often related to issues in one's waking life that were out of balance—activities, thoughts, emotions that seem muted on the one hand or in excess on the other. When mental states and behaviors leaned too far in one direction, the result was disequilibrium in one's psychological state. One's dreams often tried to correct this excess, moving one's psychic equilibrium back in the other direction, even swinging beyond the point of a healthy balance. As an illustration of this thinking, I offered an example of a person whose work life and nonworking life were out of balance.

Now, recall that in our study of dream symbolism, we included not only measures of Freudian symbols like elongated objects but also mentions of clear-cut sexuality. We found that the students who were not involved in a satisfying romantic relationship and were not having a physical relationship had more dreams that included obvious sexual content than the other students in our sample. Jung probably would have said that the students who were missing out on an important part of life were receiving compensation in their dreams, so my own data which were not obtained to test Jung's theory reported findings that could be interpreted as evidence supporting his theory.

With the followers of Jung showing little interest over the years in carrying out research that subjects Jung's theory of compensation in dreams to rigorous scrutiny, I thought it might be useful to make a suggestion or two as to how others might begin to fill this vacuum.

Let's begin with the premise supported by past research that much of the content of our dreams deals with our autobiographical memories. There is also evidence that experiences that are personally meaningful are more likely to be part of our dreams than the basic routines of daily living. Typically, we dream about interactions with other people, often important to us, activities that may be significant in our daily lives, at school, at a job, or at home. We seldom dream about brushing our teeth or combing our hair.

So far, so good from Jung's point of view. He would probably agree that the things that are pressing one in life, that are important to one's psychological makeup, are the things that enter our dreams. Here is where Jung would take us a step further. In his theory of compensation, he argued that it's the aspects of our daily life that are out of balance that would enter our dreams, not those things that are going well and are currently in psychic equilibrium. Was he right? Is this the case?

Let's focus on our *daily activities* for a moment. Make up an imaginary list in your mind of your daily activities, the ones that are important to you. Take a little time and go over some of the items. Are you spending as much time as you want to in the activity, or are you spending too much time? Or have you reached the Goldilocks' point, just about right? Now, let's draw on Jung's compensation theory and ask which of these activities should enter your dreams. Stated this way, you have the beginnings of a researchable question. I believe a straightforward reading of Jung's theory of dream compensation would suggest it would not be the activities with Goldilocks ratings that would become part of your dreams, but those with ratings of "too little" and "too much."

Now, let's take another flight of our imaginations and turn this line of thought into the outline of a study. Imagine you are teaching an introductory psychology course at a college or university. Let's say that you have 150 students who might be available for your study. You devise a questionnaire that includes a list of typical student activities. A few possible examples: use social media, get exercise, listen to music, spend time with a romantic partner, study for school courses, watch TV, and spend time with friends. In your questionnaire, you ask the students to respond to a question, such as "How much time are you spending on each activity listed?" For each question, you present the three alternatives: "not enough time," "just about the right amount of time" and "too much time." You could even use a more sensitive measure by expanding these three alternatives to five with the extreme points being "not nearly enough time" and "much too much time."

After you have collected the questionnaires, you ask the students to write down any dreams they remember when they awaken in the morning. You decide on the number of dreams you need. You examine the content of the dreams for the presence of the activities you listed in your questionnaire and probably add some broader categories as well, such as "achievement content," "social content," "sexual content," etc. As a check on the objectivity of your ratings, solicit another analyst to make these judgements independently. This will allow you to assess how much agreement there is in scoring the dreams.

If you are conversant with statistical methods, you will then set about analyzing the data to see if Jung's hypothesis is correct—that the activities marked "too little" and "too much" will be most likely to show up in the content of the dreams.

Sound interesting? Let's go on. We now have identified the incorporations of the students' activities into their dreams. If Jung is right, dreaming about these activities should have a demonstrable effect on one's satisfaction with one's waking activities. The dream should have the effect of bringing our views about these activities in daily life back to a more even keel in our overall psychological functioning.

How could we assess this? One obvious way would be to again administer the activities questionnaires to the students and see whether the ratings for the activities mentioned in the dreams have shifted in the Goldilocks direction. If that happened, it would be an additional confirmation for Jung's theory of dream compensation.

This is, of course, just a sketch of a possible study. Anyone actually trying to carry out these ideas would have to make changes and revisions. There is a lot of work to do between making an armchair proposal for a study and successfully carrying it out, but I offer this sketch both as food for thought to dream researchers and as a challenge to those who believe in Jung's ideas. Why not move beyond writing scholastic essays and carry out some replicable research to investigate whether his theory of dream compensation is supported by replicable research?

Enough of this pipe dream. Isn't that a poor choice of words? In the absence of such research, however, we can still make a few observations that bear on Jung's theory. Research does tell us that people often feel better the next morning after a good night's sleep that usually includes dreams. We also know that the activity of the amygdala, which is a center for controlling emotions, tends to be reduced the day following dreaming. Would this be more apparent for dreams that are clearly compensatory? Without serious research, we cannot really answer this question. It may well be that the reasons for a better mood following a night of dreaming may have nothing to do with compensation, but may be simply a biological phenomenon, as the proponents of the activation-synthesis hypothesis might argue. Therefore, I am inclined to offer a kind of Scotch verdict for Jung's theory of compensation. Interesting, but as yet unproven.

We cannot leave Jung without mentioning Jung's admonition to be cautious in making symbolic interpretations of objects in dreams. Clearly, Jung believed in fixed symbols. Objects in dreams could be true symbols, representing something else, not yet consciously recognized by the dreamer for what they mean. Yet, while insisting on the existence of such symbols, Jung argued at the same time that it was necessary to take into account the dreamer's "philosophical, religious and moral convictions" when making assertions about the symbolic meaning of dream images. He was admonishing people that while dream images may be true symbols, even widely recognized as such, not to jump to conclusions about what they see in the dream reports. Jung regarded "the symbol as an unknown quantity, hard to recognize and, in the last resort, never quite determinable." It is not surprising that Jung used the word "dogmatic" to describe Freud's use of dream objects and actions to indicate sexuality.

14

Dream Dictionaries

When we think of a dictionary, we usually think of a very thick book with many thousands of words arranged alphabetically from A to Z. Each word is defined to make its meaning as clear as possible and there are often several meanings given to a single word. This is not what you find in a dream dictionary. If you included the word "knife" in reporting your dreams, you would not look it up in a dream dictionary to obtain a more clear definition of what the word "knife" means. Rather, you would expect to find that a knife in a dream stands for something else, and you would probably find male sexuality. In other words, the knife would be treated as a symbol, an object representing something else, and if the "true meaning" offered in the dream dictionary were male sexuality, it would almost certainly reflect the thinking of Freud that elongated objects often represent male sexuality. In most dream dictionaries, however, the objects as well as actions in a dream could stand for a whole host of different things, not simply sex. To make this point clear, let's consider another example. Think about an overcoat, a coat, or to use a more old fashioned word, a "cloak." A dream book might tell you that a cloak in a dream might represent concealment, mystery, or a protective emotion such as warmth or love. It seems fair to think of a dream dictionary as something like a codebook in which you are told that objects or actions in your dreams are supposed to represent different objects or actions or even emotions that may not be mentioned at all in the dream.

Here are some things everyone should know about dream dictionaries:

1. Dream dictionaries are nothing new. They go far back into antiquity.

2. Dream dictionaries may offer more than one meaning for an object or action in a dream.

3. Different dream dictionaries offer different meanings for the identical dream object and action.

4. The rationale authors of dream dictionaries offer for stating that a given object or action in a dream really represents something else differs from author to author. An author might draw on the psychoanalytic ideas of Freud or especially Jung and his theories about archetypes, or look for symbols that have been used in art or literature, or draw on beliefs held in different cultures, both from the past or in the present. The author may draw on the interpretations used by past or contemporary authors of other dream dictionaries. The rationale may not be clearly specified, or simply something the author felt made sense.

5. Scientific evidence for the interpretations of dream objects and actions offered in dream dictionaries is almost entirely lacking. Some people might delete the word "almost."

Let's consider these assertions in more detail, beginning with a brief look at the history of dream dictionaries. When we began this book, I mentioned an ancient Egyptian papyrus, a dream book uncovered in the library of the scribe Kenhirkhopeshef. This book about dreams contained more than 100 dreams along with interpretations for their meanings. The interpretations suggested were not one-to-one translations from objects in the dream to something in waking life, but more like predictions made for the dreamer's future. We previously discussed the dream of eating crocodile flesh, which was thought to be a good omen, as it indicated that the dreamer would become a village official. Here are two examples from the same papyrus where dreams were thought to be bad omens. The first was uncovering one's own backside during the dream, as it foretold that the dreamer would become an orphan. The second was dreaming of making love to one's wife during daylight hours. This was also viewed as a bad omen, because it indicated that *one's god would discover one's misdeeds.*

Here is an interesting twist on the above dream interpretation. My brother Larry, an archaeologist, was excavating a site in western

Kenya, while his wife Pat was studying the history and culture of the Pokot people who lived in this area and in Uganda. When she completed her research, she wrote a book about the Pokot people, *Red-Spotted Ox: A Pokot Life*. While I was writing this chapter on dream dictionaries, I asked Pat whether she had included any mention of dreams in her book. She replied she had included only one. When talking about dreams, her informant said, "If you dream about having intercourse with your wife or husband, you must make love immediately, or something bad might happen to one of you." This was an eerie coincidence. It has been 3,000 years, more or less, since the Egyptian dream books were written, yet the idea that dreaming about sex can have negative consequences is alive and well in a tribal society that would have had little chance of ever reading a translation of the Egyptian papyrus.

As was true of the Egyptian papyrus, A. Leo Oppenheim's monograph *The Interpretation of Dreams in the Ancient Near East, with a Translation of an Assyrian Dream Book* contains brief descriptions of specific dreams that predicted future events. I provided a few examples of these in Part I.

The most well-known book about dreams from the ancient world is a multivolume work, *The Interpretation of Dreams*, by the second century AD writer Artemidorus Daldianus, also known as Artemidorus Ephesius. He lived on the west coast of Asia Minor, then part of the Roman Empire, now part of modern Turkey. He was a professional dream interpreter. While he expressed a more modern view of dream interpretation by stating that a dream interpreter needed to know something about the individual's background when interpreting the dream, he was not reluctant to offer specific meanings for different types of dreams. For example, consider dreams about teeth. Artemidorus noted that such dreams are open to many different interpretations. One interpretation was that the mouth could be viewed as a house and the teeth as inhabitants of the house. The right side of the mouth signified the men, while the left side signified the women. Teeth could also indicate the dreamer's possessions. He believed the grinders represented treasures while the canines represented objects of little value. You might be wondering if one dreamed of losing some of one's grinders if that would mean that one would lose one's possessions. Artemidorous said that could be the case. I imagine if you had recurrent dreams of losing teeth, it could portend financial ruin. A Roman citizen of the time having such dreams might have felt it prudent to seek the protection of the Roman equivalent of Chapter Seven in a bankruptcy court.

Let's move ahead in time, about 3,000 years, stopping at the year 1944 when Harry B. Weiss published a fascinating article entitled "Oneirocritica Americana" in the *Bulletin of the New York Public Library*, which describes the wide variety of dream books he found while searching libraries and antiquarian collections. Among the interesting array of dream books he found was an Oriental dream book, an Italian dream book which purported to be the greatest authority on dreams ever published, *The Complete Fortune-Teller and Dream Book*, *The Witch Doctors Illustrated Dream Book* and one written by King John of the Gypsies.

Some of the books he cited remind one of the predictions offered by the ancient Greeks and Egyptians. One book informed its readers that if they dreamed of money, they were going to experience a loss, and if they dreamed of hanging, they would become rich and respected. Another book told its readers that if they dreamed they heard a *clock strike* (you can see this is really old stuff), they would be speedily married. A fourth book stated that dreaming of knives meant you would experience lawsuits, disgrace, and marry a shrewish, unfaithful wife.

Interestingly, some books that Weiss came across linked dreams to numbers. The claim was made that if you dreamed about certain objects, you would be able to identify your lucky numbers and this could be useful if you bet on horses or played games of chance. In one book, it was said that if you dreamed about beans, your lucky numbers were 11, 18 and 72. Ready to visit that dingy store that had a numbers game going on in a secluded room in the back, bet on these numbers and make your fortune? Not so fast! A few years later another book came out that said dreams about beans meant your lucky numbers were really 25, 20 and 2. Later books changed the numbers again and then again. The old adage that there is safety in numbers certainly doesn't apply here.

Three Dream Dictionaries: Our Specimens for Analysis

Let's fast forward again and consider some current dream dictionaries and see what they have to offer. Is it old wine in new bottles? Simply more of the same with a touch of window dressing—references to ancient lore and psychoanalytic theory to add the cloak of scholarship, making the assertions advanced sound more authoritative and hence more plausible? Or could it be, wonder of wonders, the authors have tried to use the scientific method to verify their claims? Let us see.

In taking a look at current dream dictionaries, the first question that arises is what dream dictionaries should we choose. There are plenty out there. As a starting point, I took a look at the dream dictionaries that were presently being sold at an online book site. Here are some of the titles that appeared on my computer screen: *Dreamers Dictionary, Dream Dictionary, The Dream Dictionary from A to Z, Llewellyn's Complete Dictionary of Dreams: Over 1000 Dream Symbols and Their Universal Meanings, Dictionary, Dreams—Signs—Symbols: The Source Code, The Ultimate Dictionary of Dream Language, The Big Dictionary of Dreams: The Ultimate Resource for Interpreting Your Dreams*, and last but not least, a title that might bring a smile to those who share my deep skepticism about these books, *Dream Dictionary for Dummies*.

If I had felt that these books were a serious scientific enterprise, I would have been tempted to make a random search of dream dictionaries to use in this discussion. Random selection greatly reduces the biases that can come into a study, but as even a cursory examination of some of these books led me to conclude that they contained little more than unproven speculations, I decided to make my task easier by telephoning my public library and asking for the titles of a few books that were dream dictionaries or were filled with dream dictionary-like assertions. As a modest check on my own biases, I checked out the first three titles the person at the library mentioned. The three books they gave me—which we will refer to for convenience as dream books "A," "B," and "C"—are the books we will use as illustrations in our discussion. While A is not a dream dictionary, it does contain a small dictionary of common dream symbols, which is offered as a starting point for detecting personal symbols in dreams. "B" and "C" are clearly labeled dream dictionaries. The titles of these books are available in Notes and Sources.

A few thoughts before we begin. All three of the authors of these dream books seem to be influenced by the theories of Jung, particularly those relating to the collective unconscious, archetypes, and symbolism. For example, in his list of references, the author of dream book B lists works by Jung more frequently than those of any other author. In my discussion of Jung's approach to interpreting dreams, I observed that his adherents much more often published applications of his theories to various aspects of our culture than scientific studies to prove his theories. I said that I wished they had devoted more attention to proving his theories rather than simply applying them.

My gold standard for quality psychological research would be papers accepted by the refereed research journals of the American Psychological Association or other psychological journals that have similar standards in accepting research for publication. If I expressed a difference of opinion with the priorities of the followers of Jung, that difference expands to a chasm when I consider the assertions made by the authors of dream dictionaries. I applied the gold standard above to the 10-page list of references provided in dream book B and counted the number of citations from refereed research journals in psychology that studied the validity of dream symbolism. The total number? Unless my eyes missed something, the number was zero. I think this illustrates the stark contrast between what I consider convincing research on the validity of assertions made in dream dictionaries and the views held by the writers of the dream dictionaries. It is clear we look at this material through very different lenses.

Let's begin our inquiry by asking what these dream books say about some of the objects that occur very often in dreams. Using Hall and Van de Castle's table of the frequency with which different objects appeared in dreams, I chose to start with cities and streets. Remember that the dream book from ancient Assyria included predictions from dreaming about visiting different cities, so the idea is hardly new. Keeping the antiquity of the idea in mind, let's see what these three contemporary dream dictionaries say about cities and streets.

Dream book A suggests that dreaming of being lost in the city frequently represents the confusion one experiences in modern living. In addition, the book suggests that if you happened to dream of a ruined city it could be drawing your attention to aspects of your life you may have neglected, such as your relationships or your goals in life. As I read this, I wondered how many people actually dreamed of ruined cities. In this country, I would suspect very few. Perhaps an archeologist like my brother. I'll have to ask him.

Then, I looked through the alphabetical listings in dream book B and found that the author asserted that if the setting of your dream is a city or town, this may deal with choices. The choices included those you might make about relationships, activities, and direction. This seems a lot to hang on being in the city in your dream. Getting lost in a town indicates you are unsure about your place in society.

I found no listings for cities or towns in the index for dream book C and moved on.

How about streets? I saw little in dream book A relating to streets and I did not find anything in the index of dream book B pertaining to streets. However, the author of dream book B had a lot to say about roads. He asserted that roads represent both your direction in life and approach to it. The interpretation of dreams about roads in dream book C seemed somewhat similar in that all roads were said to be symbols of one's movement through life and that dreams about urban streets may relate to one's previous social experiences in making stops along streets.

A few observations before we proceed further. Consider the two assertions made about dreams about being lost in a city. One author looks at such dreams as coping with the confusions of modern living while the other talks about such dreams as having to do with your place in society. The two authors present different ideas about the meaning of such dreams. Who is right? How could you tell? Maybe neither is right. Now, suppose instead of relying on these two books, you look through six or seven more books with dream interpretations and came up with four or five more ideas about dreams of getting lost in cities. Where are you now?

The interpretation that being on a street is really an encapsulation of your life experiences rather than simply supplying the necessity to be *somewhere* in your dream strikes me as a very long reach. If you are not allowed to be on a street in your dream without having to accept the interpretation that you are really watching your life go by seems to me really pushing things.

Since we have been talking about streets, let's imagine that our dream places us in an automobile, perhaps driving along one of those roads where our lives are said to go flashing by.

To provide context, remember that Hall and Van de Castle's data tell us that cars are common objects in dreams. While dream book A is relatively quiet about cars, dream book B devotes several pages to the meaning of cars in our dreams. In his opening sentence about the meaning of car dreams, the author of dream book B presents his view that such dreams are about drives that motivate you. Some of the possibilities he listed are your sex drive, your ambition, and your sense of failure. Further along in the discussion, the author develops the idea that other aspects of our lives such as decision-making and relationships may also be involved—all of these significant aspects of human behavior seemingly stemming from *a dream about a car*.

Then, the author offered a qualification. It could be that a dream about a car is simply a dream about a car, something that we utilize in our everyday life. I read this as a recognition that a car dream may be about none of the ideas he previously listed.

I found this qualification refreshing and was happy to see it, because it was one of the rare times in my perusal of dream dictionaries that I found myself in agreement with an author. After this brief qualification, the discussion about car dreams continued, and in time, interpretations were offered for dreaming about the various parts of a car. Here is a small sample of the interpretations offered: the body of the car might be your own body, the trunk of the car could represent a number of possibilities such as memories, anxieties and thoughts, while the tires could represent the skills and attitudes enabling you to smooth out the rough patches you encounter in life. What happens if a tire goes flat? According to the author, this might mean injuries or physical problems.

You may be wondering whether there is any validity to these assertions, such as a dream about cars may represent your sex drive. Using the standards of evidence that I briefly mentioned and will describe in detail later, my answer is I am not aware of any studies that prove these assertions.

Dream book C has less to say about cars but is nonetheless interesting. It tells us that the dreams involving a car reveal things about the journey of your life, and if your car has crashed or broken down, you may have a feeling that your life has stopped moving forward. It also tells us that the color of your car in a dream may have some deeper meaning, but at this point, doesn't specify what that meaning is. I wonder if a creative person at an advertising agency might extend this idea to cars. Who knows? Tapping into deeper meanings for colors (if they do exist) might raise sales.

Let's take an object less often reported in dreams, the mirror. The mirror is interesting because we often use it to look at ourselves, and doing this in a dream might plausibly be interpreted as looking more deeply at ourselves. What, if anything, might this reveal? I was not only curious what the writers of dream dictionaries might make of this; I will also throw in a notion offered about mirror dreams in the ancient Egyptian papyrus, and for good measure, the ideas of some learned psychoanalysts.

Dream book A suggests that looking into a mirror in a dream may be a sign of narcissism. If the face you see in the mirror is a strange one rather than your own, this could indicate an identity crisis.

Dream book B suggests that looking into a mirror in a dream indicates a concern about how other people feel about you. As was the case with dream book A, it can also be an indicator of self-love (a synonym for narcissism). It may also indicate anxiety about aging and suggests you are looking into your unconscious.

Dream book C also suggests that looking into a mirror in your dreams may indicate that you are seeing what the unconscious mind is seeing. The author suggests that if you are comfortable with what you see in the mirror, this indicates that the way you view yourself in life is real. However, if the image seems out of kilter, it indicates you are not being authentic, perhaps hiding something from someone. The use of a mirror in the dream was also viewed as comparable to the concept of cause and effect, a kind of parallel to using a tool to figure out things. Interestingly, the notion is advanced that the meaning of looking into a mirror depends on the size of the mirror—a handheld mirror might be presenting an intimate glimpse of one's self, while if the dream mirror is full length, it suggests the need to take a more holistic picture of your present life situation.

I would like to comment about two ideas about mirror dreams in the interpretations offered by the writers of the three dream dictionaries, that looking into a mirror in your dreams may be looking into your unconscious mind and that it also may be a sign of self-love or narcissism. Looking into one's unconscious is a rather fanciful idea and I am not sure how one would go about it in reality. Freud wrote that dreams were the Royal Road to the unconscious and he believed that free associating to dreams and then interpreting the associations with the aid of his theory of dreams would be a way of probing into one's unconscious. We have looked for research evidence validating Freud's approach to dream analysis and found it very limited. While his method of dream analysis is the product of a brilliant mind, without such supporting evidence, the results of such analysis are suspect.

In contrast, narcissism is a much more down to earth idea, visible in everyday life, well described clinically and measurable by psychological tests. Narcissism is usually defined by such terms as self-love, egoism, vanity, and a fascination with one's self, particularly with one's own body.

The term is based on the story from Greek mythology about a man named Narcissus who was much admired for his beauty. One version of the myth says that Narcissus was the son of the river god Cephissus and the nymph Liriope. Narcissus was said to be a very proud man who rejected the people who loved him. His pride in his own beauty angered one of the gods, who lured Narcissus to a pool where he saw his own reflection in the water and fell in love with himself. Narcissus was unable to move away from the sight of his own reflection and continued to stare at it until he died. One version of the story said that he committed suicide because there was no way he could fulfill the love of himself.

When you think about it, the word "narcissism" in part overlaps the concept of self-esteem. Being narcissistic is much like being at the top of a scale of self-esteem. It is like self-esteem gone through the roof. In fact, there is a diagnosis in the American Psychiatric Association's *Diagnostic and Statistical Manual of Mental Disorders* which includes a category called "Narcissistic Personality Disorder." It is characterized by a grandiose sense of self-importance, preoccupation with ideas of unlimited success, exhibitionistic needs for attention and admiration and an extreme focus on one's self. You may have encountered people like that in your daily experiences or you may have seen them on television. At the other end of the self-esteem continuum, you'll find people who feel they are completely unworthy, undeserving, and often are very depressed.

But let's be careful about jumping to the conclusion that looking into a mirror during our daily activities is a sign of narcissism. I suppose if one spends a great deal of time admiring one's self in front of a mirror that could well be true, but the ordinary use of a mirror during the day hardly strikes me as narcissism. Typically, when a man uses a mirror in the morning as he shaves, it is to protect himself from cutting his face repeatedly rather than a sign of narcissism. Looking one's best can be a significant advantage in our society, both on the job and in the pursuit and maintenance of romantic relationships. Checking oneself in the mirror before a social engagement is a useful tool in daily living in our society, hardly an indicator of a personality disorder.

Now, let's turn to dreams. What if you dream of looking into a mirror? Is that a sign of narcissism? My opinion on this one way or the other would be just as dubious as the assertions made by the authors of dream dictionaries that indeed it is so. The idea could be stated as a research question, however, and I would challenge the proponents of dream dictionaries to carry out such research. Here is a relatively simple way to do it. There are psychological tests for narcissism. Give the tests to people who say they have had "looking into the mirror dreams" and compare their responses on the scale with a matched control group of people who say they have never had such dreams. See whether the mirror dreamers have significantly higher scores on the narcissism scale. If they do, I will gladly concede the issue and tip my hat to the researchers. If they don't, well, maybe the researchers could admit that this is just one more example of an unproven conjecture.

Let's go back to what the writer of the ancient papyrus had to say about looking into a mirror in one's dreams. The translation of the papyrus states that seeing one's face in the mirror is a bad omen, meaning a new life. I don't see the logic of this, myself. If you have a good life to begin with, the idea of starting a new life, or even moving into new directions, might seem like a bad idea and the dream omen might seem like a dire warning. But what if the reverse were true, and things are not going well? The omen's forecast of a new opportunity might not seem all that bad. Anyway, I must admit that I am puzzled by this. If you find this a bit puzzling, too, try figuring out what two contemporary psychoanalysts have to say about what they call "the common mirror dream."

The common mirror dream (CMD) is one in which the dreamer looks into a mirror and sees himself or herself distorted. The image is recognizable but clearly altered. According to the authors of an article published in the *Journal of the American Psychoanalytic Association*, such dreams are not very common. From what I gather, they have to do with the dreamer's relationship with his mother. After working with some patients with this problem, the authors noted that all of their patients felt

> enjoined by the mother (in most cases with the father's collusion) not to see and regard her clearly and not to be an accurately reflecting mirror for her. (I am not exactly sure what that means but let us read on for clarification.) The intensity with which the maternal injunction against accurate visual perception and evaluation was feared was an important distinguishing feature in our patients with CMDs. The essence of the CMD has been hypothesized to be a reciprocal, reverberating, visual—exhibitionistic dyad representative of the mother—child relationship. The dream mirror may represent the wish that the analyst—mother counter a feared parental injunction against accurate visual perception and evaluation so as to correct the distorted perceptions of self and objects and provide visual affirmation of the value and integrity of the self-representation. For some patients, defense against the dangers of castration and loss of maternal love was accomplished by the mirror mechanisms of magically transforming images in the mirror, the ease of creating illusion in the mirror, and a fetishistic mechanism of visually reintrojecting a phallic symbol from the mirror.

If you are uninitiated in psychoanalytic terminology, you might find the above unintelligible. While I am highly critical of dream dictionaries, the ones that I have looked at are at least easy to read and some of the dictionaries (e.g., dream book C) are very well written. To help you gain some understanding of what the folks above are saying about common mirror dreams, here are a few modest translations:

Re-introjection. Introjection is usually thought of as a defense mechanism in which one incorporates the views and values of another person into oneself.

Fetishistic. A fetish is a sexual attachment to any inanimate object.

Phallic symbols. Think back to my study on Freud's theory of sexual symbols. Think about the list of male symbols. They are supposed to represent the penis.

Castration anxiety. According to Freud, boys are supposed to experience anxiety about castration in their psychosexual development.

Does any of this help? I hope so.

My reading of what they are saying is, that based on their clinical work with their patients, they believe that a man who looks into a mirror and sees a distorted image of himself may be experiencing severe conflicts with his mother and that his mother has made it clear that she doesn't want her son to perceive her correctly. The son is uncomfortable with the situation and hopes that the psychoanalyst he is seeing will correct the situation, allowing him to look at the situation involving both his mother and himself correctly.

Now comes the harder part. The interpretation of these unusual dreams plunges headlong into Freudian theory. As far as I can determine, the authors believe that when some of these men see a distorted image of themselves in the mirror, this image can have the characteristics of a phallic symbol. These men may form an attachment to this image (a sexual fetish) and reincorporate this into themselves and in this process protect themselves both against the loss of their mothers' love and against castration.

Do I believe that there is a simpler, more plausible explanation for distorted mirror dreams? I certainly hope so. I'm not a Freudian and the explanation offered above sounds very far out to me. For many modest distortions, such as when people look into the mirror and see themselves looking poorly and less attractive than others would see them, the distortion may have more to do with the tremendous importance we place on physical attractiveness in our culture than it has to do with mother-son conflicts. The heavy cultural pressure many people feel to look attractive could certainly create distortions in perceptions. An explanation along these lines would easily apply to people who are very depressed and often look at themselves in negative terms whether or not they look into a mirror. While this alternative explanation is just an impromptu thought and may not be relevant to the kind of distortions the psychoanalysts observed, I find myself very skeptical of the analysis they offered.

Returning to dream dictionaries, in my view, any time you read that an object or action in a dream means something else than what it appears to be, it is a reach. Some reaches in dream dictionaries seem longer than others and may make you wonder. Here are a few translations of objects in dreams that strike me as long reaches. Let's see what you think. Here are a few examples from dream book A:

A spider in a dream may represent a devouring mother who consumes her children by her possessiveness and her capability to arouse guilt.

Have you had a hair in your dream? It could indicate vanity.

Ever dreamed of women's shoes? They could stand for dominant female sexuality.

Let's turn to dream book B:

A house in a dream almost always refers to yourself. This includes your body and aspects of personality.

The attic of a house may stand for your mind.

The chimney may represent the birth canal.

Now, let's peruse dream Book C:

Blue is said to be a color representing communication, so if your dream is predominantly in blue, you should consider issues of communication in interpreting your dream.

Have you ever dreamed about birds? In dream book C, birds are said not only to be messengers, but when you have a dream that includes birds, the dream may also be trying to inform you that there is already information flying about in your consciousness that is ready to be used.

Have you ever dreamed of swimming? Dream book C tells us

that your dream may be about something other than the act of swimming. The dream may be about reviewing the emotions you experienced in your life, i.e., your emotional journey. How you swim in the pool can yield information about how you are doing in the emotional aspects of your life.

You may be scratching your head and asking if there really is proof that dreaming about a chimney means that you are really dreaming about the birth canal or dreaming about swimming is really a representation of an emotional journey. From my standpoint, the answer is very clear. *No.* Then, you may well ask, how can the authors of dream dictionaries make these claims? My answer is that they appear to have a different standard of evidence from the one I have tried to present in this book. Let's pursue these differences in more detail.

The Standards of Evidence

I have not hidden the fact that I am deeply skeptical of dream dictionaries. Why am I so? It all has to do with the standards of evidence. In studying dreams, I believe it is important to strive as much as possible to approach the standards of science, and this certainly applies to the assertions made in dream dictionaries. When a chimney in a dream is said to represent a birth canal rather than a chimney, I want it to be proven. The authors of dream dictionaries make little effort to do this in a manner that satisfies even the minimum requirements of the scientific method. They appear to be satisfied with their own set of standards, which, in my judgment, fall far short of those of science.

Because the writers of dream dictionaries eschew validating their speculations with serious scientific research, they are open to the same criticisms I made about the psychoanalytic community. I would really like to see the people following both traditions make a serious effort to tighten their standards of evidence. Recall the study of the common mirror dream reported in the *Journal of the American Psychoanalytic Association*. While offering their elaborate interpretation of these dreams, the authors also stated, "Our data failed to confirm many of the hypotheses of previous contributors as to specific symbolic meanings of the dream mirror." What were their data? Not a controlled study. It was simply working with a relatively small sample of patients and coming to a conclusion. Having said this, I appreciate the fact that they called their conclusion a *hypothesis*. They recognized the difference between believing something is true and having solid proof that it is so. In making their assertions, the writers of dream dictionaries do not always make this difference clear to their readers. Too often, their assertions appear to be stated as facts.

Belief is not a substitute for science. Centuries ago, most people believed that the sun revolved around the earth. It was very easy to believe this. After all, it was common sense. However, it just wasn't right. The pioneering astronomer Galileo was kept under house arrest by the Vatican for stating this belief was incorrect, and the monk Bruno was burned at the stake for a similar offense.

While members of the psychoanalytic community share a predilection with the purveyors of dream dictionaries not to test their theories by carrying out replicable scientific research, I certainly do not mean to lump the two communities together in the way they do things and approach dream interpretation. The practice of psychoanalysis often involves a very long, protracted process in which the analyst has nearly hour-long sessions with the patient. The therapist has ample opportunity to get to know the patient's personality and patterns of thinking and behavior intimately. When the patient describes his or her dreams, the therapist may find himself or herself in a good position to integrate the dream into his deep knowledge of the patient's ways of thinking and behaving. I would have more confidence in the value of what they might offer if they are able to free themselves from a dogmatic insistence on orthodox Freudian notions and rely on their knowledge of their patients rather than on what strikes me as questionable theory. In any event, I would hope that they make it clear to their patients that the interpretations offered about dreams are hypotheses, not facts. This injunction should apply to every type of psychotherapist, not just psychoanalysts.

For their part, the writers of dream dictionaries come to their conclusions in a variety of ways—reading scholarly studies, examining myths and traditions, reading other people's works about the meanings of dreams, relying on psychoanalytic theories, particularly the work of Jung, interpreting the dreams of other people, or making semantic analyses—extrapolating from the common uses of objects and letting their knowledge and imagination offer surmises about the potential meanings of objects and actions that are reported in dreams. This sort of semantic analysis may be fine for developing ideas, but, as I have said, too often, the authors pass off these surmises as *facts, not speculations.* Too often, we read categorical statements that this is a symbol of such and such. I do not doubt that the authors may hold these beliefs as a matter of deep conviction, but in the final analysis, a conviction is one thing; proof is another.

Two quick additional points. Number one: while the thinking of Jung is an important source for the writers of dream dictionaries, Jung made it very clear that one should not interpret information from dreams as symbols in a this-means-that automatic fashion. He argued that it was important to take into account the person's background in making any interpretations. Moreover, he regarded dream symbols as hard to recognize and never quite determinable. One wonders how this admonition squares with many of the assertions made in dream dictionaries. Number two: the use of semantic analysis to establish alternative meanings for dream objects and actions can lead to any number of conclusions. Take the color blue. Remember, dream book C stated that blue was the color of communication. Dream book B offers entirely different meanings to the color blue such as being depressed and experiencing religious feelings. If different authors come to different conclusions about the same words, where are you? Does one have to take a vote among dream dictionaries to see which interpretation comes up most frequently?

There is no shortage of books making claims about the real meaning of objects and actions reported in dreams. What there are shortages of are science-based studies proving that what they say is true.

What, then, is the alternative? I think it is very clear from my gold standard for psychological research that the alternative is to apply the rigors of the scientific method as much as it is possible to do so. Psychologists trained in the academic discipline of psychology and holding a Ph.D. are schooled to apply the methods of scientific inquiry in their research as far as it is possible. Here are some things they try to do:

1. The hypothesis to be tested should be stated in a clear and testable way. For example, "People who state that they have had dreams of looking into a mirror will have higher scores on measures of narcissism than people who do not state that they have had dreams of looking into a mirror."

2. The development and use of methods of data gathering should be as objective as possible. Two people who look independently at the data collected from a person should be able to interpret the data in essentially the same way.

3. The methods used to test the hypothesis should allow for quantification. Numbers allow for precise comparison.

4. Comparisons between groups of people and the establishment of correlations between measures should be subjected to statistical analysis. These analyses can tell us whether the findings observed are merely chance findings or whether they indicate real differences.

5. The findings should be written up clearly and conservatively and in such a way that another investigator can repeat the study and see whether the results can be replicated.

6. The study should be reported in a refereed journal. It can be reviewed for competence by one's peers in the field.

Carrying out research on a problem such as dream symbolism is admittedly not an easy one, but it can be done and should be done. My own study looking at the validity of Freud's view of dream symbols is an example that shows that dream symbolism is a researchable subject.

Do I really think there is a chance that the proponents of dream dictionaries will carry out serious research to prove their assertions? Research that approximates, as far as possible, the standards of science? I certainly hope that some of them will consider the idea, but I'm not optimistic. Dream dictionaries will continue to be published and widely read. I would like to see more cautionary statements included in them such as the thoughtful ones expressed in dream books A and C, respectively, that dreaming is a personal experience unique to the individual and that there is no single correct approach to interpreting one's dreams.

15

The Dream Incident Technique: An Experimental Approach to Using Dream Associations

Both Freud and Jung relied on associations to dreams as an essential tool in probing into the meaning of dreams. Freud used a free-wheeling approach to dream associations by asking his patients to offer their associations to almost any part of the dream narrative. Moreover, he asked them to relate whatever came into their minds, holding nothing back, no matter how irrelevant and trivial it seemed to them. Jung took a much more focused approach to dream associations by picking and choosing those aspects of the dream he felt were important and then asking probing questions to try to find out what these associations meant to the dreamer. What these two approaches to dream associations had in common was that, in the final analysis, they relied on the skill, experience, and creativity of the dream interpreter—the psychoanalyst—who was conducting and orchestrating the dream analysis. Depending on the skill and theoretical leanings of the analyst, there was no guarantee that different analysts would extract the same associations, and even less so that he or she would arrive at the same conclusions when the dream analysis was completed.

On the first standard where we suggested a more scientific approach in assessing an individual's dreams, objectivity, Freud's approach of free association to dreams does not fare at all well. It is very subjective. Nor does it lead to any kind of precise measurement, either qualitative or quantitative. Jung's approach to obtaining dream associations is tighter, but it still falls far short of objective assessment.

You may ask, if this is the case, why bother with the association approach at all? The answer is asking for associations to dreams is the best approach we have to retrieve the memories that are the basis of our dreams. Research suggests that dreams are based on our experiences, and particularly those experiences that are significant. With the proper questions, such associations can zero in on the experiences of our lives that seem to relate to our dreams. Moreover, it is possible to develop methods of dream analysis using dream associations that pass muster with all the criteria for a more scientific approach that we have just laid out. The experimental method that demonstrates this is called the Dream Incident Technique (DIT). In the form that it was developed, it is a research instrument rather than a clinical one, but the materials used in its construction can provide a more objective guide for people in analyzing their dreams.

A little personal background may be in order here. As I began to look through the articles I had written about the DIT, I began to reflect on what prompted me to develop the technique and to carry out a program of research with it. When I thought about this, I realized that the genesis of my work in this area occurred back when I took my first college courses in psychology. At the time, the work of B.F. Skinner and his Skinner box were all the rage in psychology and particularly in the psychology department where I was studying. The department seemed devoted to promoting Skinner's ideas and following in his footsteps. For those of you who are unfamiliar with Skinner, let me say that he was the ultimate behaviorist and the only data that he seemed to trust came from studying the behavior of white rats in a simple device that was called the Skinner box. The box was simplicity itself. There was a lever the rat could press and if the rat pressed the lever it would be rewarded—*reinforced* as the term began to be used—by receiving a pellet of food. I am not sure what was in the pellet but the rat must have liked it, because it would keep pressing the bar to get one of the pellets to eat. In the experimental psychology course I was taking, another student and I were given temporary stewardship of a Skinner box and a white rat. We were asked to run through a series of studies in which the rat was given various feeding schedules, such as being reinforced every time it pressed the lever, or perhaps every third or fifth time it pressed the lever. The measure in these studies was how many times the rat would continue to press the bar when it was not reinforced at all. This was called *resistance to extinction.*

The procedure was very orderly and scientific and the results became predictable. However, it did not take me long to conclude that the results were essentially meaningless in terms of understanding the complexities of human behavior.

My conclusion that this was not the way to go in psychology was buttressed when I attended a lecture in another psychology class where the professor said that he could tell whether psychotherapy for humans that was directive was better or worse than psychotherapy that was nondirective by comparing the behavior of two rats in their Skinner boxes. When I raised my hand and said that this was nonsense, I noticed a few heads nodding in the class, but these were the days in which students did not argue with professors, but simply wrote down what they said. The professor responded to me by making a rather labored defense of the idea that one could extrapolate from the lever-pressing of rats in a box to the relative effectiveness of two forms of psychotherapy with human beings. I had the unhappy feeling that he really believed this. Today, of course, different forms of psychotherapy are compared for their effectiveness in rigorous studies on patients, but this is today and my story fortunately is about a long-ago yesterday.

Anyway, not long after these events, I was invited to the home of a college friend whose mother was a practicing psychoanalyst. When she asked me about my psychology courses and I related what I was being taught, she shook her head in disbelief and said I needed better than that and made arrangements for me to enroll in a psychoanalytic institute where I joined a small seminar of psychiatrists and psychiatric social workers who were studying Freud's lectures. While I found Freud's ideas fascinating and relevant, I could find no sign of any serious proof for what he had written, so I felt that I had gone from the frying pan, working with rats in Skinner boxes in an orderly, scientific manner, which I believed told me next to nothing about human beings, to the fire, learning some fascinating ideas about human personality that had little basis in science. When I finished the course, I left with the notion that someday I would like to take the interesting ideas of Freud and see if it were possible to test their validity in a more scientific way. This possibility came to pass some years later when I was on the staff of a university department of psychiatry with time to conduct research, working with colleagues who were interested in what I was doing. One of the fruits of this collaboration was the development of the DIT, a technique that used dream associations as the basic data, followed by objective assessment of what the associations meant to the dreamer, leading to quantitative scores that could be used to test hypotheses.

Like any other technique of dream analysis, the DIT involves assumptions. The first assumption, similar to those of Freud and Jung as well the dream dictionaries, is that dreams are not meaningless activities of the brain but present information to the dreamer that is relevant and meaningful. In this respect, I disagree with the views of Hobson and the proponents of the activation-synthesis hypothesis who regard dreams as a biological phenomenon essentially devoid of meaning, or, to put it bluntly, simply noise.

The second assumption is that a substantial part of the content of dreams is based on our memories of personal experiences, what researchers have more recently called autobiographical memories.

Assumption three states that these memories often relate to unresolved issues the individual faces in his or her waking life. These issues are assumed to be a source of tension for the individual, perhaps only mild tension or perhaps more significant, but in any event, the issues have some motivational and emotional significance and probably exert some influence on the way the individual feels and behaves in his or her current life.

The fourth assumption is that dream associations offer a path toward identifying the nature of these unresolved issues and the tensions they produced. From the variety of associations that might be offered to dreams, I decided to narrow the range of possibilities to specific incidents the dreamer recalled when thinking about the dream. I believed that using such incidents would not only be a good choice for zeroing in on real life experiences but would also be ideal in simplifying and standardizing the technique when in actual use.

Finally, the technique assumes that these real life incidents can be examined by the dreamer and rated using the kind of scales that psychologists have developed for conventional personality tests. When I talk about recalling incidents that are associated with dreams and postulate that these incidents are associated with areas of unresolved tensions in one's life, the reader might comment that this sounds something like Jung's theory of compensation. Indeed, it does. Jung postulated that issues in one's life that were not attended to would tend to find their way to our dreams. But I was not in the least influenced by Jung in developing the DIT because I developed the instrument years before I ever read a word of Jung. It is not uncommon for people interested in a subject to follow similar paths independently of one another.

Let's look at the procedure of the DIT in detail.

In using the Dream Incident Technique, the person carries out three activities in sequence. The first task is to report a dream. The second is to think of past experiences (incidents) that come to mind when reflecting on the dream. The third is to rate each incident using a specially devised checklist which asks about different kinds of motivations.

The instructions to report a dream are simple enough. People are instructed that as soon as they awaken in the morning to try to recall whether they had a dream during the night. If they recalled a dream, they were asked to write it down immediately, as dream recall often fades very quickly as the day goes on. In the event that several dreams were remembered, the instructions were to write down the dream that was remembered best.

The instructions for thinking of incidents that were associated with the dream were more detailed. The users of the DIT were told to read their dreams over, think about what they read for a while, and then go back and read over the sentences of the dream very slowly, one at a time. Then they were asked to relax and let their minds drift freely over the content of the dream.

As they thought about the dreams, they were told to ask themselves if any incidents that actually happened to them came to mind. These could be incidents from any time in their lives but it must be something that they, themselves, were involved in. There could be other participants in the incidents such as family members, friends or acquaintances, or it might be that the incident involved no one other than themselves. As these incidents occurred to them, they were asked to write them down. They were instructed to write down as many different incidents as occurred to them up to a total of seven incidents for the dream. The incidents were written on separate sheets of paper.

If they had trouble thinking about incidents, they were asked to try reading the dream over once again, but if 15 minutes went by and no incidents occurred, they were told to stop. They were reminded to write down only specific events, something that happened at a given time and place.

Let's turn our attention to the rating scales. I could have asked the people taking the DIT to rate these incidents in a number of ways. For example, I could have asked them to indicate what feelings they might have had during the incident. For example, were they happy? Depressed? Angry? This might have worked out well with interesting results. I thought about it, but instead chose to ask them to think about the motivations they may have had both before and during the incident. What were they trying to do? What were their hopes, desires and wishes? In making this decision, I certainly was influenced by Freud. Remember, Freud said that dreams were really a fulfillment of wishes. For Freud, of course, these wishes were primarily sexual, but there are many very important wishes that people have that have little or nothing to do with sex—to achieve something, to be with friends, to seek adventure, to help or protect others, or to seek help or protection yourself, etc. What I wanted to know was what motivations people were experiencing during these real life incidents—incidents that were brought to mind as they thought about their dreams. Because these motivations were dream-related, I assumed they were colored by tension.

In choosing a list of motivations to use in the DIT, I drew on the classic writings of the distinguished psychologist Henry Murray. Murray developed a list of psychological needs to evaluate the responses to the projective technique he devised called the Thematic Apperception Test. You may remember that we cited the test in our discussion of nightmares. The test consisted of presenting a set of interesting pictures to people to look at, such as a boy looking at a violin. The person was then asked to make up a story based on what he or she saw in the picture. The stories could then be inspected for the presence of a variety of motivations such as achievement, aggression, dominance, and so on. I used 12 of the motivations from Murray's list, the three examples above (achievement, aggression, and dominance) plus affection, autonomy, adventure, sex, social recognition, nurturance and play. I also included two motivations of avoidance—guilt and embarrassment. Then I wrote down a series of items that reflected these motivations. For example, for achievement, I included the item, "To excel; to attain a high standard." For autonomy, I included, "To be independent; to be free to do things my own way"; for adventure, "To go to new places; to do new things"; for dominance, "To have my ideas, my way of doing things, prevail"; and for sex, "To kiss, to pet—to make love."

At this point, people using the DIT were asked to ignore the dreams and concentrate only on the incidents. They were asked to consider each incident one at a time using a checklist of 35 items similar to the examples given above. The written instructions were, "Think about the incident you described above. If you felt any of these hopes or wishes either before or during any part of the incident please indicate by checking YES. If you did not feel the hope or desire listed, check NO. If you are uncertain, check ?."

You can see that the DIT is a complex procedure involving several stages, so let's pause for a moment to make sure we are all on the same page. Perhaps the easiest way to do that is to imagine that you are one of the students who was using the procedure in one of my research studies, or perhaps you might be thinking that you would like to try the procedure yourself to see if it is helpful in understanding your dreams. In either case, imagine that you have just awakened in the morning, and with paper and pen by your bedside, you immediately write down the dream you best remembered from your night's sleep. Then, imagine that you read and reread the dream, paying attention to each of the sentences in the dream and then began to write down incidents that came into your mind as you thought about the dream. Now, you study the checklist of items (below) and ask yourself whether any of these ideas, hopes and wishes occurred to you either before or during the incident. You check YES if they did, or NO if they did not and leave a ? if you're not certain.

Here is a sample (the first 10 items) from the checklist of 35 items as they appeared in our studies using the DIT. For readers interested in using the DIT to analyze their own dreams, the complete list of 35 items along with the identification of the 12 motivations they measure is in Notes and Sources.

1. To lead or direct the activity.
2. To seek adventure.
3. To protect, care for.
4. To excel; to attain a high standard.
5. To have my ideas, my way of doing things, prevail.
6. To behave in such a way that I would not feel guilty about it afterward.
7. To feel close to someone of the opposite sex.
8. To resist attempts by others to tell me what to do.
9. To behave in a way that I would later feel was completely responsible.
10. To physically attack, to hit, to strike, to slap.

Let us imagine that you have worked your way through the entire list of 35 items, checking off the ones that you were experiencing before and during each dream-related incident. Let's say that when you looked through the items that you had checked off for the various incidents, you often saw the item, "to excel; to attain a high standard." If you checked this and similar items, such as, "To accomplish, strive, achieve," you can see that you were motivated to achieve something during these incidents. Some of the other items you may have checked among the 35 were transparent too, such as, "To experience sexual pleasure," and "To kiss, to pet—to make love." The motivations in these real life incidents clearly included the desire to engage in sexual activities.

Scores for the DIT were hypothesized to indicate current levels of tension for these different motivations. The scores were calculated in the following manner. The denominator for a given score was the number of potential items in a given area (e.g., achievement) that could be checked "yes." This denominator varied with the number of incidents reported. If there were three items sampling an area of motivation (e.g., achievement) and four incidents were described and rated, the denominator would be twelve. If five incidents were described and rated, the denominator would be fifteen. The numerator was the number of items checked "yes." The incident score was a simple percentage, the number of items checked "yes" over the possible number.

These scores made it possible to carry out many types of research, such as relating DIT scores to other measures of personality and testing a variety of hypotheses, and we were able to do so by applying tests to evaluate the statistical significance of the results.

While the DIT was designed as a research instrument, you can see how it may be useful as a guide to thinking about your dreams and what they might be suggesting to you about unresolved issues in your life. I would like to illustrate this possibility first and then describe some of the scientific research using the technique.

Imagine once again that you have written down what you dreamed about, written down the real life incidents that were suggested by the dream and then assessed them using the checklist of statements describing different types of motives. I think you can see how looking at the statements you checked "Yes" would tell you something about the wishes, aims, and desires that you may have experienced before and during your dream-associated incidents. You have evaluated the experiences that may have given rise to your dream. Now, one dream is a very small sample of your dreams over the weeks, and like Jung, I would caution you not to put too much store in considering a single dream. However, if you try this procedure on a number of dreams and you begin to find some common threads in the way you assess these incidents, then you might begin to form a hypothesis that the motives you continue to identify are issues in your daily life that are unresolved and are worth paying more attention to. I would urge you to treat your initial surmises as educated guesses to be refined or discarded as experience dictates, not something set in stone.

My book *The Psychology of Dreams* contains a number of illustrations of how the DIT seems to have shed light on unresolved issues in the lives of some college students who participated in my early research. Because some of the dreams reported are lengthy and the incidents described in detail, I will only offer two abbreviated versions here, knowing that the full case studies are readily available in my earlier book.

When Betty took the DIT, she was a 21-year-old college student in her junior year. She was an active person with a well-rounded life, had several close friends, participated in a number of sports and frequently dated a special boyfriend. Their relationship was warm and affectionate and included sex. Her grades in school were average. Scores on personality tests revealed nothing problematic about her. One noteworthy thing she revealed on a questionnaire was that while she did report that she was occasionally irritated by other people, she rarely expressed anger openly and avoided arguments.

The events of her dream were complex, taking place in several settings with unclear time changes. In the beginning of the dream, she was in her old house asking her mother for some special kind of paper needed to write up her physiology laboratory report. Then she found herself walking along her elementary school playground where she met three bridesmaids dressed in pink crepe as well as the maid of honor who was dressed in a mulberry dress. As she approached a bookstore, she once again encountered the bridesmaids and then found that she had trouble with her own dress, constantly stepping on the hem. One of her sister's friends appeared and called her "corpulent." The dream moved along as she went into a theater with a man who at first glance seemed to be her brother. While they were in the theater, they got into a dispute as her brother had possession of her purse, which contained his driver's license, and he refused to return it to her. Feeling angry, she left him, walked down the hall and found herself no longer in the theater but in the old house she was in at the beginning of the dream. Still angry, she discussed her brother with her mother. Then, he was getting married and neither her nor her parents were at the wedding. Then she was again alone with her brother, who it turned out was no longer her brother but someone else. His new wife had gone away and the man was suggesting that they go to bed together and she laughed at him.

As I reread this case many years after I first included it in *The Psychology of Dreams*, I found it even more interesting than when I read it the first time. The many changes in scene are atypical when one recalls Hall and Van de Castle's data, and the time sequence of her brother's marriage is confusing. The change of identity of her brother into a promiscuous man eager to sleep with her is interesting, and the fact that he held her purse with his own driver's license inside of it would surely arouse Freud's interest, as in his view, the sexual symbolism would be unmistakable. But let us now depart from this intriguing manifest content and consider the real life incidents which came to mind as she thought about her dream.

The first incident recalled was about her pink nightgown. On the previous night, she wanted to put on her nightgown and discovered that she hadn't mended it. She said it had been a while since she had torn it and she could not remember when it was torn.

The second incident described the time earlier in the year when a friend of the family got married and the maid of honor was someone she had gone to school with many years ago. The dress she wore at the wedding was pale green or pale lavender.

The third incident was about the paper she used to write up a lab report earlier in the day. She didn't like the color of the paper. It was an unpleasant avocado green.

The fourth incident concerned a friend's husband who had picked her up after work one evening and took her out for a cocktail. While taking her home, he drove to a remote spot with a nice view and told her that she was sexy and she would make a good marriage partner. He tried to kiss her and run his hands over her. She became irritated and told him to take her home.

The fifth incident involved a trip to the city with her sister and the sister's friend who in the dream had called her corpulent. The final incident occurred years ago when she was wearing a candlelight-colored ball gown. She attracted a lot of attention.

Ratings for these incidents showed three important differences with the way she was behaving in real life. Her dream incidents for aggression were high, suggesting unresolved problems. In her daily behavior, she was unwilling or unable to express anger but her dream incident scores for aggression were high. The same pattern seemed true for autonomy. In her waking life, she had many friends and was actively involved socially, while her dream incidents scores suggested she might be experiencing conflict about her level of freedom. This might reflect issues living at home with her parents and her siblings, or perhaps conflicts with her friends. Finally, her grades at the university were only fair, while her DIT scores for achievement were high. This suggests she might not be giving enough time to her schoolwork and may be aware of this.

This analysis suggests that concerns about personal freedom, anger and achievement are motivations that appear to relate to her dream content, perhaps suggesting tensions in her life, which, on the surface, appears to be running smoothly. Once again, it is important to remember that these are initial impressions, a preliminary hypothesis based on a single dream. Whether we would find additional evidence to support this hypothesis or would be inclined to disregard it depends on looking at a larger sample of her dreams and the incidents that are associated with it.

Our second illustration, once again taken from *The Psychology of Dreams*, concerns a young woman named Margaret. Margaret was a 21-year-old college senior when she completed the tasks involved in using the DIT. Like Betty, Margaret was an average student academically. Unlike Betty, Margaret had a strong interest in artistic activities, writing poetry, etching, and working with graphic arts. Margaret was very much an outdoor type of person who loved to camp, climb and hike. She had little interest in more sedentary social activities like parties, and she saw herself as a person with strong autonomous needs. However, she did have a steady boyfriend with a warm and affectionate relationship on both sides.

The two dreams that I obtained from Margaret were very different. The first dream seemed to reflect her experiences in the outdoors. The second dream, which I will summarize here, brought up other aspects of her personality and took her away from her usual satisfying pursuits.

She dreamed that a very close friend, a hiking companion probably belonging to the hiking club that she was a member of, had been killed in a mountain climbing accident. In the dream, she saw him standing in a field covered with snow just before he tried to begin a difficult climb to a hard-to-reach peak. She left the scene because of some sort of errand or she would have climbed with him. When she was back home, she read in the newspaper about the accident and then a friend asked her if she knew that Tony was dead. She reacted by prolonged sobbing. Now she had the responsibility of managing his estate, including the distribution of his property to his friends and relatives. She felt this was an unwelcome, unpleasant task. However, when she went to his residence to begin this task, she found that the building she went to was not Tony's residence at all, but the home of a friend who was a painter. The manager of the building would not let her enter the door but she still had to carry out her responsibility. Everyone she tried to contact about doing this was hostile to her, telling her that she had no right to do what she was doing, and she ended up sleeping in the back of Tony's car. Meanwhile, many of his rock climbing friends were scheming how to steal the equipment he had used in his rock climbing. She was very frightened and saddened by this experience and went to sleep in the dream. Margaret noted that the dream experience was extremely real and for a time afterward she actually believed that Tony was dead until she heard from him during the following week.

Margaret recalled three incidents as she thought about her dream. The first incident was really a compilation of deaths that occurred in her family and among her friends. She recounted the loss of her grandfather, an aunt who she dearly loved, her father, her stepfather, and a close friend who died in an auto accident in which she was present. All of these deaths occurred during the last six years. As a consequence, she said she felt shaky about developing close relationships.

The second incident concerned the setting of the residence that she mentioned in the dream. She said that the residence was associated with another person who was not in the dream, rather it was the home of a close friend who had moved to New York. Margaret had not seen that person since he moved, nor did she expect to, but she was curious about how he was doing in graduate school. She thought he was a very talented painter and would like to see him succeed. She reiterated that they had been close friends.

The third incident was about the person involved in the dream. He had asked her to be the executor of his will. This was not something she really wanted to do, but she felt she should do it because he was a friend. If she had to carry out this responsibility, she would have to deal with his family, and she felt they wouldn't approve of her and certainly she should not be the person to carry out his instructions.

The ratings that she gave to these incidents highlighted nurturant feelings—the desire to care for and protect others. These motives were not apparent in the personality tests that we gave her. Rather, she saw herself as an autonomous, adventurous, and artistic person, far from a nurturant caregiver. Yet her anguish about all of these recent deaths brought out these latent needs and the feeling she must be responsible to do what was asked of her. The ambivalence she had developed about forming close relationships is another interesting aspect that comes out in the dream analysis, and it would be interesting to see how she deals with this in her waking life and whether it is further highlighted in her dreams.

Finally, here is a recent dream of my own, one that relates to the writing of this book:

I was in a very large library. I wasn't sure which library I was in, as it seemed unfamiliar. I knew it was not the Library of Congress, which I had used for scholarly work many years ago when I lived nearby on Capitol Hill in Washington, D.C. I assumed it must be the National Library of Medicine in Bethesda, Maryland, but it looked unfamiliar to me. There was a balcony in the library I did not remember and there were desktop computers on every desk rather than piles of medical journals. I noticed that there were few people in the library. I was not actually looking up things but I was using the desktop computer to put down some thoughts about this very book that you are reading now. Suddenly the room darkened. It was no longer possible to do any work. Then, someone brought candles to put on the desks where people were working, including one for my desk, but I found the light inadequate and I couldn't work. I woke up at this point.

A few minutes later, I was asleep again and it was a continuation of the dream, but this time there was no desktop computer, only a small tablet of the sort that I was totally unfamiliar with and couldn't figure out how to use. Sharon, who is much more fluent than I am with computers and often gets me out of jams when I use computers, was sitting across from me, patiently waiting for me to finish my work. I was trying to do the best that I could but I could not access the material I had typed before on the desktop computer. I wanted to print what I had written, but there was no printer available, and the more I fumbled about with the tablet-like device, the more frustrated I got, nearly to the point of tears. Meanwhile, a slice of delicious-looking cake appeared on the desk. Finally, I heard loud voices coming from above on the balcony. I saw a television camera and an interview was going on. It was all very loud and distracting. I woke up again.

Incident 1. I used the National Library of Medicine several times when I was writing *The Psychology of Dreams*. The library was uncrowded as it was in my current dream but devoid of any computers, and I ordered journals using slips of paper. It took a lot of time but everything went very smoothly.

Incident 2. I had called our local library system earlier in the day to see if it had been able to find a monograph on dream interpretation in the ancient Near East. I had put in a request for it two months ago and I had heard nothing. The woman I talked to was very helpful and said she would try to expedite matters.

Incident 3. In thinking about the cake, Sharon told me that she had gone to a birthday luncheon that day and had bought a roasted chicken and a plate of fruit as her contribution. On the way into the party, without asking, an acquaintance tore off a piece of the chicken to eat, somewhat spoiling the appearance of her contribution to the lunch. She was not happy about it.

Incident 4. As I thought about things that were not working well in the dream, I recalled that earlier in the day I had opened a compartment in the refrigerator that stored vegetables, only to see a piece of plastic that supported the compartment snap off. I thought about trying to glue the plastic part back into the refrigerator but realized it would probably be futile. The compartment filled with vegetables was unlikely to hold under the stress.

Incident 5. Another thought about libraries. I remembered when I wrote *The Psychology of Dreams*, I needed to find a reference, once again dealing with dream interpretations in the ancient world. After many phone calls, I finally located the reference in the central library of Washington, D.C., now named the Martin Luther King Library. I was able to get into the city, find the manuscript and take the notes needed to include in my book.

Incident 6. In a showing of silent films, I saw Charlie Chaplin's memorable *Modern Times* in which his little tramp character fumbled haplessly to cope with the machinery of what was then modern times.

In looking through the DIT items, I could see a number of patterns: in these incidents I was definitely trying to achieve something, there was clear evidence of avoiding embarrassment, some evidence of offering support or nurturance and a need for help. When I initially began to develop the DIT years ago and tried it out on myself, I also had high scores for achievement, but there were also high scores for autonomy, a need to try to do things my own way. Achievement continued to be a major focus over the years, which is reflected in the score of books I have written, but the emphasis on autonomy may have subsided. Increased emphasis on helping and being helped certainly reflects the patterns of life as one grows older. There are no DIT items for confusion or difficulties of adaptation to changing environments. If they had been included in the DIT's panoply of motives, I would have checked them both. With the pace of change in today's world, particularly in the use of technology, I suspect that I am not the only one who has encountered this problem. One may either smile at this or shake one's head in consternation. I suspect smiling is the better way.

Let's now turn to the DIT as a research instrument. As you read through the cases using the DIT, you could probably see the possibilities for obtaining revealing data about the individual and the issues he or she might be dealing with. You may have wondered about different aspects of the procedure and you might have some questions you would like to ask. I thought we might begin our discussion of the research undertaken with the DIT by trying to answer a few questions that might have occurred to you and other readers. For example, you may have wondered about the number of incidents people wrote about in response to their dreams. Remember, we allowed our subjects to write down incidents up to a total of seven. What was the average number? In one of our samples, we looked at this question and found that the average (mean) number of incidents reported was 4.2.

I think you would agree with me that most of the incidents reported in our illustrations seem well connected to the manifest dream content. You may have wondered whether sometimes incidents are brought up that bear only a slim connection to the dream. The answer is yes. For example, in one of my early research reports about the DIT, I mentioned a subject who had a dream in which she took a typing test for a possible job. I have to note that this dream was reported in the days before personal computers were widely available to the general public. Even desktop computers were a novelty, and the mobile devices of today did not exist. The typing test in the dream turned out to be bizarre, as the dreamer was instructed to put a pencil in the typewriter and type on the pencil rather than on paper. In the dream, she became flustered and performed horribly. Interestingly, one of the dream related incidents she reported had nothing whatsoever to do with her performance on the typewriter. Rather, it had to do with her singing performances. She said she loved to sing, and when she was a teenager she sang several solos in school. Unfortunately, while singing, she would get very nervous and would sing very poorly. The dream and the incidents it caused her to recall suggested problems relating to stage fright that may still be unresolved.

It would be a very interesting research question to ask whether incidents that very clearly related to the dream or incidents like the one above which are more remotely associated to the dream are more revealing about personality and unresolved issues. I think I know what Jung would have said about this question but I am not so sure what Freud would have said. What an interesting debate that would have been.

Dreams can be very different from one night to another. Even though the dreams may be different, would there be any consistency between the DIT scores from one dream to another? The answer is yes, there is some consistency for most of the motives assessed. The highest levels of consistency in DIT scores were for our needs dealing with self-assertiveness such as autonomy, dominance, social recognition, achievement, and, interestingly, for the measure of avoidance of embarrassment. Among the lowest levels of consistency were the scores for affection and nurturance. Tensions relating to the kinder, gentler side of our personality needs seem to ebb and flow. A low level of consistency was also found for DIT sex scores. Possible explanations for inconsistency in incident measures across dreams for sex include fluctuations in biological sex drive, the availability of a partner, and changes in one's general emotional state.

I found the pattern of more consistency in the self-assertive motivations than in the softer, more loving ones interesting and wondered what you might make of it. My first thought is that it might reflect our culture, which, like most modern societies, has a strong competitive thread running through it. Competition to succeed, if not excel, in so many important areas of life, in school, in sports, in jobs, may create more chronic tensions in our lives than the needs to be kind, affectionate, and cooperative. Perhaps we would find a different pattern in less competitive societies, such as the Israeli kibbutz, isolated communities where people must depend on one another, and in people living in religious orders such as a convent. But perhaps the more assertive measures in the DIT that seem more chronic in our dream-related incidents are simply part of the human condition. I once carried out a study in a very large home for the indigent aged and found that the people judged by the nursing staff to be the most aggressive tended to live the longest.

I have criticized both the psychoanalytic community and the writers of dream dictionaries for not carrying out serious, replicable research that offers evidence supporting their assertions about dreams, so it now is only fair that I tell you about some of the research I carried out with the DIT, aided by my colleagues Roland Tanck and Arnold Meyersburg. Both Roland and Arnold were friends and colleagues working at George Washington University, Roland in the Counseling Center and Department of Psychology, and Arnold, like myself, in the Department of Psychiatry in the medical school.

In the development of the DIT, I assumed that the scores for the various motives indicated levels of tension about these motives. If you had a high score for DIT achievement, it suggested that you had high levels of tension about achievement. You may or may not be fully aware of this tension in your waking life, but a DIT score for achievement indicated that this may be an issue in your life that has not been fully resolved. Is this really the case? Here is a study I carried out that suggests this is indeed the case.

Imagine that you are a person with ambitions to move ahead in life and achieve success, and then imagine that you are having anything but success. One would guess that this should cause you some disquiet, even unhappiness. The need to achieve and the failure to do so could well be an unresolved issue producing tension. Taking this idea as a starting point, I hypothesized that people with high motivations for achievement who were not having success would have higher DIT scores for achievement than other people (people who simply had low levels of achievement need in the first place, or if they had high levels, were successful in their undertakings).

What I did was give out a personality test that included a widely used measure of need for achievement to students at the University of California at Berkeley and then also obtained measures of their grade point achievements. From this sample, I selected students with *high achievement needs* whose great point averages were in the *lower half of the sample.* I compared these students with high achievement needs and low performance with the remaining students in the sample using the DIT score for achievement. The prediction that the students who wanted to achieve but were not doing so would have higher DIT scores for achievement was born out. The results were statistically significant, indicating this was not likely to be a chance finding.

Replication of studies is an important benchmark for science, and I repeated the study when I was at George Washington University not once, but twice. In both studies, students with high achievement needs and lower grades scored higher than the rest of students in the sample, and in one of the two replication studies, the results once again were statistically significant.

Here is a study with a clear-cut hypothesis, tested, producing a predicted result, and then replicated.

Let's talk about a second study using the DIT. When I had completed the study at Berkeley, I took a position at the George Washington University School of Medicine in the Department of Psychiatry. It was a very good position for me because my teaching and clinical duties were modest and I had plenty of time to do what I most enjoyed, research. Early in my stay there, I became interested in psychosomatic medicine. One of the important issues in the field was the influence of emotions on physical symptoms. There was much interest in the possible effects of emotions such as anxiety and hostility on the development of physical symptoms. I wrote a monograph on the relation of personality and psychosomatic illness, which summarized much of the thinking and scientific research in this area, and then set about carrying out research of my own. I was particularly interested in finding out whether people who were experiencing high levels of tension and were unable to effectively deal with the problems that were causing this tension would be more likely to experience a range of common symptoms, such as headaches, dizziness, fatigue, and gastrointestinal problems. I thought the DIT, which purported to measure unresolved tensions, might be a good instrument to investigate this question.

Because my research partner Roland Tanck taught very large classes at the university, we had an opportunity to recruit research subjects. When we thought about tensions in university age students, three of the areas of tension believed to be assessed by the DIT—achievement, autonomy, and sex—seemed clearly relevant in the everyday lives of the students. Accordingly, we formulated our hypotheses about these three issues. In regard to coping with ongoing problems, I could not find a measure of the ability to cope with stress that was really satisfactory, so I used an approximation of this, a measure of what is called *ego strength*. People scoring high on this measure would not be expected to fall apart and become unglued when they had to deal with their problems but would persevere. They would try to do the best they could.

To test the three hypotheses, we needed to obtain a reliable assessment of the student's current physical complaints, and rather than rely on a simple questionnaire, I recruited a group of senior medical students, well trained to take histories and physicals, to interview each of the students and obtain this information. I did not ask the medical students to do physicals but worked with them to use a standardized procedure to take medical histories. The medical students were very interested in the project and did a careful job. Our predictions that students who had high levels of tension in the Dream Incident Technique's areas of achievement, autonomy, and sex and had relatively low ego strength would report higher levels of physical symptoms was confirmed in all three instances. When considered alone without any reference to ego strength, the DIT scores for autonomy and sex were positively related to physical complaints.

So here we have evidence that measures of tension derived from incidents associated to dreams were related to the level of physical symptoms the students reported. Therefore, when I consider theories like the activation-synthesis hypothesis, which asserts that dreams are primarily random phenomena of the brain, my response is, "I don't think so."

Let's move on to the third study, which is in many ways the most interesting. I think you will find the results tantalizing.

Bear with me for a moment and imagine you are sitting comfortably in a chair reading from a page of a science fiction novel. "As the spaceship descended through the thick layer of clouds above the forbidden planet, the captain looked at the screen in front of him and his eyes widened in amazement at the sight of the mile-high buildings that loomed below on the planetary surface that was presumed to be uninhabitable." Interesting, you might think. The book might be an entreating read, but what does it have to do with the DIT? If you narrow your reading focus to the significant words, "his eyes widened in amazement," the answer is everything.

Think of your pupils for a minute, the dark part of your eyes that often dilate when you look at something that excites your interest in a positive way. Many years ago researchers conducted experiments in which they found a tendency for males to experience pupillary dilation when looking at pictures of nude females while females showed the same reaction when viewing pictures of partially clothed males—and of pictures of babies. When experimenters showed subjects pictures that would seem to most of us very unpleasant, such as dead bodies, they found that rather than dilating, pupils tended to contract, as if the subject were trying to shut off the offending stimuli.

One of the interesting features of pupillary reactions was that the reactions seemed to be involuntary, out of the subject's control, and even out of the subject's awareness. These features interested me greatly because of some obvious similarities to the nature of dreams. Dreams are usually involuntary, out of the subject's control and awareness, and memory for dreams is usually fleeting. When my colleague Roland Tanck told me that he was developing hardware to measure pupillary dilations and contractions, we discussed the similarities to dreams and then developed a research design that would enable us to correlate data from two very different techniques of data-gathering, pupillary dilation and the DIT—both offering access to phenomena that appeared largely involuntary, occurred with diminished awareness and seemingly tapped the realm of unconscious behavior. In general, we wanted to know whether people with high DIT scores in a certain area (e.g., achievement) would evidence greater dilations to scenes representing that area than people with lower DIT scores in that area.

As we worked out the details of our research design, we decided we needed to include another measure of the personality needs measured by the DIT, this one a conventional self-report measure of personality without any links to dreams or any other clear links to what might be unconscious materials. We found a personality test that met our needs precisely. The test was the *Edwards Personal Preference Schedule*, a carefully constructed self-report personality test that also used the Murray system for describing personality needs.

Having resolved the problem of providing a control measure of personality needs using the Murray system, the Edwards Personal Preference Schedule, we turned our attention to the remaining major challenges to carrying out our research. The first challenge was to have a workable pupilometer, an apparatus that would measure the students' pupils as they watched our visual stimuli. The second was to have a set of scenes to portray on a screen that would be clear depictions of the needs measured by both the DIT and the Edwards test. With my own mechanical aptitude barely measurable, I was glad to leave the first task to Roland who was both good at and enjoyed tinkering with gadgets. He devised a contraption that when compared to the high-tech devices now used in pharmaceutical laboratories would be something like comparing the Wright brothers' first airplanes with a modern supersonic jet. To process the data that were obtained was extremely labor-intensive, but the system did work— at least most of the time, although we did lose a little of the data. Meanwhile, I turned my attention to devising the scenes that the subjects would look at.

In doing research, there is usually a lot of what can be sheer drudgery, but there are also times when the work involved can be sheer fun. I think everyone must have had a dream at some time of making a movie, and this was as near as I ever got to it. Constructing a short film strip was one of the most enjoyable tasks I ever did during my research years. After drawing up a number of possible scenes, we obtained the cooperation of the university's Speech and Drama Department and it obliged us by recruiting student actors, providing and using film-making equipment, and providing us amateurs with real expertise. We shot a number of scenes depicting six of the areas measured both by the DIT and the Edwards test. These were achievement, nurturance, affiliation, sex, aggression and dominance. For example, one of the scenes depicting achievement showed a woman receiving a trophy. A scene for nurturance showed a mother cuddling a small child. In depicting aggression, we had a woman slap another woman in the face. This was something we had to get right on the first take. I can't imagine a "take one," "take two," "take three," and "take four" without provoking a rebellion among the actors. Fortunately, the first slap filmed was fine. In retrospect, I really should have given the woman who was slapped a box of chocolates. Portraying sex was also a delicate issue. The task was to make our couple look like they were engaged in passionate sexual activity when in fact they were doing nothing of the sort, but these students were very good actors and we managed to pull it off.

In our research design, we wanted to find out whether the self-report measure of personality we selected, the Edwards test, would relate to pupillary reactions or whether it was only the DIT measure with links to dream associations that would relate to these reactions. Our college students were first given the Edwards test and then asked to report their dreams and follow the instructions of the DIT to write down and then rate their dream-associated incidents.

Now, let's see what we found. A technical observation first. Each area of tension measured by the DIT was depicted on the film by two or three separate scenes. If a person's pupils dilated to one of the scenes portraying a given area, such as achievement, would he or she tend to dilate to the other scenes depicting that area as well? The answer was yes. For several of the areas portrayed such as sex, achievement, and aggression, this was the case, indicating some consistency of pupillary responses within a given area.

With this assurance, we then moved on to another important preliminary question. Were the pupillary reactions made without awareness? We thought this would be the case, but now was the time to find out whether this was true. When we correlated the individual's self-report of feeling stimulated by each of the video scenes with their actual pupillary dilations to these scenes, we found that the results were not statistically significant for all 16 pictures in our film. The results were strictly chance. The relationships were no better than one might pick using a table of random numbers. The individual's pupillary reactions to the video scenes appeared to be independent of his or her own reports of stimulation. Our data suggested that there was no awareness of this body reaction.

Now our test of the hypothesis. None of our control measures of personality needs, the scales of the *Edwards Personal Preference Schedule*, showed a significant relationship to pupillary dilations. The correlations were quite small; the statistical tests clearly indicated only chance relationships. In contrast, three of the six hypothesized relationships between the DIT scales and pupillary dilation were confirmed. DIT scores for tensions relating to sex, achievement and nurturance were all positively related to pupillary dilation to the scenes portraying these themes in our film. Interestingly, for the DIT score indicating tensions relating to *aggression*, the correlation was negative and very high. Instead of dilating to aggressive scenes, people with high tension levels relating to aggression tended to contract their pupils. As we have shown, this appears to have taken place without awareness. As this was contrary to what we hypothesized, we can only wonder and speculate about it. One thought that comes to mind was the experiments of others reporting about pupillary reactions to unpleasant stimuli such as pictures of dead bodies. Recall that the subjects in these studies tended to contract their pupils when looking at such pictures. Could this be key to what happened in our study? We really don't know, but in the paper we published, we noted that aggression has a more negative social connotation than the other motivations we looked at. It might be that persons having problems with aggressive feelings experienced high levels of discomfort when viewing the aggressive stimuli and attempted to avoid the stimuli. If this were the case, pupillary contraction might represent an immediate avoidance response—an interesting possibility that led us to examine similar phenomena.

One of the first things we looked at was the array of studies that were carried out under the rubric of *perceptual defense.* The idea of a mechanism of perceptual defense when one is exposed to stimuli that are personally threatening was a subject that elicited considerable research beginning in the 1950s and continued for many years. Researchers would present the stimuli they were interested in to subjects using a tachistoscope, a device that could present material at exposure speeds too fast to be recognizable and then at slower speeds which might be barely recognizable, and as exposure speeds slowed, were more easily recognizable. The researcher could then determine at what speed of presentation the subjects were able to accurately identify the stimuli. Subjects whose ability to identify stimuli accurately at difficult to visualize levels were thought to be perceptually vigilant while subjects who were only able to fully identify the stimuli at slower, easy to recognize speeds were thought to exhibit perceptual defense. Perceptual defense was thought to be a kind of internal process that warded off full recognition of what the person was viewing.

Here is an example of such a study carried out in 1994 using women who were in the third or fourth month of pregnancy. The women were asked to identify a number of pictures including some relating to pregnancy, sexuality and a father image. The pictures were presented initially for very brief durations with the time exposure gradually increasing. The women's scores for perceptual defense were recorded. Some months after the women delivered their children, they were assessed for levels of depression using a standard depression rating scale. Analysis showed that the women who would eventually become depressed after childbirth had been slower in identifying the stimuli relating to pregnancy, sexuality and the father image during their third or fourth month of pregnancy than the women who did not become depressed. What we see here is a tendency for perceptual defense or avoidance to be predictive of the later depression that sometimes ensues in the months following childbirth.

One more observation comes to mind as I think about our findings suggesting the possibility of a defensive reaction to aggressive stimuli among our students with high DIT aggression scores. This relates to some research of my own not directly connected to the DIT. Roland Tanck and I had collected some dream reports from his college classes in another study. Our research design included the collection of two sets of dream reports from the students with an interval of time in between. We collected the first set of dream reports under routine conditions at the school and then waited for the projected interval of time to elapse. During the waiting period, events took place on the national scene that caused great turmoil and unrest in the cities. There were riots in the streets. The last thing on our minds was the research we were doing, and, like everyone else, we were upset and on edge as riots occurred in our own community. Military equipment had been placed not far from the university hospital where I had my office. When things finally settled down, we collected the second group of dream reports that had been written down during and shortly after the riots. We decided to see whether there might have been an increase in aggression in the dream reports that were written during and after the riots. Much to our surprise, there was a definite decline in the aggressive content of the dreams that were reported during and after the turbulence. When we published these observations, we were candid to state that we had no clear idea why this had happened, but suggested it might have some relationship to the studies that were then being conducted on perceptual defense. I find this all very tantalizing but I remain unclear what happened and why.

The basic take-away from our third study using the DIT is that the DIT, which has its roots in the unconscious activity of our dreams, was predictive of our pupillary reactions to appropriate experimental stimuli—and that these pupillary reactions appear to be out of our control and conscious awareness. This strikes me as a very intriguing finding. Also keep in mind that this association was not found for a very well-constructed self-report measure of personality needs that had no known links to unconscious activities.

The results of the several studies using the DIT are consistent with the ideas that (1) dreams have important meaning and are not simply random events of our brain and that (2) dream associations that are clearly linked to incidents in one's life represent a source of important information about ourselves. Are our results conclusive evidence that the study of dream associations provide important links to our personality? No. Science does not work that way. The studies using the DIT are exploratory, a probe into the unknown, rather than anything like the final word.

16

Some Final Thoughts
About Dreams

While speculations about the nature and meaning of dreams can take us back into our ancient past, our current thinking about dreams really dates back to the 19th century and particularly to the clinically-based theories of Sigmund Freud. Freud's ideas became very popular in the 20th century and still influence many people when they are asked about dreams and their interpretations. In the latter part of the 20th century, our understanding of dreams began to change when science entered the picture, primarily with the groundbreaking studies on the biological level of the relationship between rapid eye movement and dreaming, and, to a lesser extent, with an increase in the rigor of psychological studies of dreaming, which have given us a much more detailed and clearer picture of what we dream about and the factors that influence our dream reports.

The studies relating REM to dreaming provide convincing evidence that most people dream even if they have little or no recall for their dreams. However, the link between REM and dreaming is not one-to-one; people experience dreams or dream-like phenomena during periods in which REM does not occur. Therefore, the biological link to dreaming is more complicated then we might have first assumed. Biological probing into the nature of dreaming continues and scientists have found that REM activity and NREM activity appear to be linked to activation and deactivation of different areas of the brain. Whether the ambiguities that remain in the effort to link dreaming to brain activation will eventually be resolved and ultimately reveal a fully coherent picture of the relationship of our dream experience and brain activity remains to be seen. A reasonably complete picture of how eye movement activity and brain activity are related may be within our grasp while a more reductionist attempt to link brain activity with precision to our subjective dream reports may prove to be an elusive goal, beyond the reach of our current technology.

On the psychological level, the studies of Hall and Van de Castle and others have told us much about what people dream about. The statistics are not only useful for providing a context for researchers in evaluating the data they obtain in dream studies; they may also act as a guidepost for individuals in assessing whether their dreams are much like those of other people or tend to be unusual and even unique. Because these studies are now dated, there is a need for a large-scale contemporary study to look at what people report in their dreams today. Changes in the way we do things in our culture over the years have been pronounced and they might be reflected in the content of our dreams. In addition to providing basic research about our dreams, psychological studies have also yielded practical techniques to help us diminish the intensity of the very unpleasant dreams we call nightmares.

While few would argue that research has told us much about the nature of dreams and dreaming, there remains some dispute about the meaning and implications of the biological findings relating to dreaming. As we have said, some dream theorists who have reflected on these data believe that dreaming can be best understood as a routine function of the brain, producing images that are of little significance in terms of providing insight about the dreamer or self-understanding. Taken to the extreme, this view of dreaming would suggest that our dreams would offer little more aid to self-understanding than the mechanisms of breathing or digestion.

Therapists who do psychotherapy and use dream interpretation as a method of probing into the thinking and personality of their patients would argue that this position is simply incorrect. They could point to the agreement of their patients when he or she points out an interpretation of the dream. They could also point out that their patients have improved following psychotherapy. As a therapist who used dream interpretation readily during therapeutic sessions, I am sympathetic to this point of view. My patients found the process of reporting their dreams to me and my questions about their dreams and the interpretations that grew out of the questions were helpful and enlightening, and, like most people who undergo psychotherapy, most of my patients felt better after a course of therapy. However, I must point out two important caveats. The first is that the therapist has the socially defined role as the expert, and people are more inclined to accept the views of experts, whether it's the assessment of a general about military strategy, a physician when it's about your health, or a lawyer about a divorce settlement. The status of an expert can have an important influence on the willingness of a person to believe what he or she has been told. The second caveat, one that I have emphasized in this book, is that most people who undergo psychotherapy obtain benefits by simply being in the process of psychotherapy because of the nonspecific effects of being in treatment. Talking about troubling issues to someone who really listens and seems interested in understanding and helping a person can have salutary effects regardless of the nature of the therapy employed. The bottom line is that the convictions of a therapist are not hard evidence, only informed opinions. When I say that I believe my use of dream interpretation with my patients was helpful in therapy and this buttressed my conviction that dreams are indeed meaningful sources of information, this belief falls far short of being hard evidence proving that it is so.

Better evidence that dreams are indeed meaningful and can be useful for increasing self-understanding comes from research indicating that a large part of the content of dreams is autobiographical memory. While sometimes dreams appear fantastic, much more often dreams are about things that actually happened in our lives. If dreams were completely random, artificial firings of the brain, why would we dream of things that happened to us, or at least associated in our minds to things that happened to us? Using real events, real context, real people, and real words, even when thrown together somewhat haphazardly, our dreams very often reflect events that have happened in our lives both recently and in the past.

If the prevalence of autobiographical memory in dreams suggests that dream content is not simply random noise in the brain, then the fact of recurrent dreams adds further weight to the conclusion that dream content is more than a quixotic display of images. Dreams can have significant meaning for the individual.

Then there are the studies carried out with the DIT, which has shown in several different contexts that it is possible to make verifiable predictions from dream associations to other aspects of the individual's life. I am fully comfortable with the idea that dreams are meaningful and that understanding the meaning of the dreams is possible, but here is a point I must stress. While the application of science to the nature of dreaming has led to a vast increase in our understanding of the nature of dreaming, this increase in reliable knowledge has not extended nearly as well to the interpretation of dreams. Here we have little more than the theories of the psychoanalytic movement, which are shrouded in the jargon of psychoanalysis and buttressed by the perception of expertise, but offer little in the way of replicable scientific evidence, and the speculations offered by dream dictionaries, which offer little more hard evidence of their validity than the speculations offered in the ancient papyrus of long-ago Egypt. The people practicing dream interpretation in both communities seem comfortable with what they are currently doing, and as long as enough people take their assertions seriously, there is little incentive for them to undertake the difficult research to prove their assertions. I suspect a large number of people will continue to read dream dictionaries without asking the very basic question, "Is this true?"

So, in regard to dream interpretation, my advice is to maintain a healthy degree of skepticism about what anyone tells you about the meaning of your dreams, whether the advice comes from people in prestigious occupations, from people who write dream dictionaries, or from friends or family. In this list I would certainly include the suggestions I will now offer, for the unhappy fact is no one really knows with a high degree of certainty what any given dream means. While trying to gain insight about yourself from thinking about your dreams is a worthwhile endeavor, be cautious when you draw your own conclusions. Here are my thoughts:

1. When you decide to explore the use of your dreams in an effort to increase your self-understanding, it makes good sense that you try to get as accurate a report of your dreams as you can. Without having the equipment of a sleep laboratory, you will have to rely on low-tech procedures. If you wake up during the night during or after a dream or in the morning after a night's sleep and you recall a dream, the thing to do is to either write down your dream memories as soon as you are able to do so, or, if you have a recording device, record your memories immediately. The word immediately is the operative word, as we all know that memory for dreams scatters rapidly and may disappear completely in a few minutes. A simple diversion such as talking to someone can eradicate memories from the dream. Place that recorder or pen and notebook by your bedside and use it quickly.

2. Jung put it well: don't rely on a single dream. I couldn't agree more. The content of your dreams may vary considerably from night to night, and while the associations you may make to your dreams and the way you evaluate your associations will have some carry-over from dream to dream as we showed in our studies using the DIT, there are likely to be unique features both in the content of your dreams, in your associations, and in the way you assess your associations. See what you can make out of several dreams before venturing a hypothesis as to what they may suggest to you.

3. Try associating to your dreams. Freud had a brilliant inspiration when he developed his free association technique. Like Jung, however, I don't think every thought or object in a dream is necessarily of great significance. Freud loved to make a big deal out of what would appear to be insignificant details in developing his theories, but I would put my money on real-life events that are suggested by the dream. Studies using the DIT suggest that the real-life events are important and I would suggest you use the instruction of the DIT to obtain a reasonable number of dream-related incidents, whether they are relatively recent events or those of your distant past. I do not know which are more important, but I would certainly include both.

4. Examine and consider each of your associations, asking your-

self what you were trying to do, accomplish, experience or avoid during these events. What were your emotions? How did you feel during and after the event? Using the items in the DIT may be a good starting point for you.

5. If there are objects in your dreams that strike you as symbols, something else other than what the dream tells you they are, be careful in jumping to a conclusion that these objects really stand for something else. Universal meanings attributed to objects in your dreams other than that what they purport to be is largely unproven theory.

6. When you have collected a fair number of dreams, dream associations, and you have reviewed and thought about these associations, then it is time to advance some very tentative hypotheses about how these dream-related data relate to your active, waking life. Do you see any guidance stemming from your dream analyses about things you might want to do differently? If you are a person who is doing pretty well in life, not feeling any more than the usual ups and downs that come with living and adjusting in a world that presents its share of major and minor challenges—and importantly, are not overwhelmed by emotional issues such as anxiety or depression, just a pretty normal sort of gal or guy—I suspect your dream analyses will probably not suggest the need for major changes. You may find that there are some things that your work with your dreams and associations suggest to you that could improve the quality of your life, and if so it might be worth trying to do a few things differently. Perhaps take a few small steps, make a few changes, and look at things in your life a little bit differently. The small steps might help. If your life becomes a little sweeter, then your dream analyses might have been helpful to you. That's not bad.

7. If you are experiencing serious discomfort in life, then I wouldn't attempt dream analysis on your own. If you are in emotional trouble, it is no time to be an amateur and engage in what could be distorted self-analysis. A good therapist may be able to help you explore the various facets of your life in a caring and objective manner. That would be the path I would follow.

If this caveat suggests that I am a cautious person, you are right. Dreams are a fascinating part of the human experience, but use your common sense and avoid the pitfalls of entertaining unproven speculations about the meaning of your dreams that could be misleading.

Notes and Sources

Notes and sources are listed in roughly the order they are discussed in the text.

Part I: What Science Tells Us About Our Dreams

Definitions of dreaming.

The definition offered is not from a single dictionary but represents a composite view from various dictionaries. The title of Stevenson's poem is "I Dreamed of Forest Alleys Fair."

Some dream researchers have refined the benchmark that dreams have story-like characterstics, viewing dreams more like a succession of fragments of stories, not usually a full-length narrative. See, for example, Montangero, J. (2012). Dreams are narrative simulations of autobiographical episodes, not stories or scripts: A review. *Dreaming*, 22, 157–172.

Dreams during daytime naps.

Carr, M., & Nielsen, T. (2015). Daydreams and nap dreams: Content comparisons. *Consciousness and Cognition*, 36, 196–205.

Daydreams.

Giambra, L. M. (1979–1980). Sex differences in daydreaming and related mental activity from the late teens to the early nineties. *International Journal of Aging and Human Development*, 10, 1–34.

Rapid eye movements (REM).

There are many sources dating back to the original discoveries of the relation of REM to dreaming, e.g., William Dement's pioneering doctoral thesis *The Physiology of Dreaming* (1958) carried out at the University of Chicago. For an easily accessible current source, see *Brain Basics: Understanding Sleep* offered by the National Institute of Neurological Disorders and Stroke.

Discovery of REM.

A very readable source is a book by William Dement, *Some Must Watch While Some Must Sleep*. (1974). San Francisco: Freeman.

REM deprivation.

Dement, W. (1960). The effect of dream deprivation: The need for a certain amount of dreaming each night is suggested by recent experiments. *Science*, 131, 1705–1707.

Oudiette, D., et al. (2012). Dreaming without REM sleep. *Consciousness and Cognition*, 21, 1129–1140.

Rosales-Legarde, A., et al. (2012). Enhanced emotional reactivity after selective REM sleep deprivation in humans: An fMRI study. *Frontiers in Behavioral Neuroscience*, 6, 25.

REM sleep behavior disorder.

The case study was reported by Ingravallo, F., et al. (2010). Injurious REM sleep behaviour disorder in narcolepsy with cataplexy contributing to criminal proceedings and divorce. *Sleep Medicine*, 11, 950–952.

Olson, E. J., et al. (2000). Rapid eye movement sleep behaviour disorder: Demographic, clinical and laboratory findings in 93 cases. *Brain: A Journal of Neurology*, 123, 331–339.

Abad, V. C., & Guilleminault, C. (2003). Diagnosis and treatment of sleep disorders: A brief review for clinicians. *Dialogues in Clinical Neuroscience*, 5, 371–388.

Arnulf, L. (2010). [REM sleep behavior disorder: an overt access to motor and cognitive control during sleep]. *Revue Neurologique*, 166, 785–792.

Lecair-Visonneau, L., et al. (2010). Do the eyes scan dream images during rapid eye movement sleep? Evidence from the rapid eye movement sleep behavior disorder model. *Brain: A Journal of Neurology*, 110, 1737–1746.

Tachibana, N. (2009). [Historical overview of REM behavior disorder in relation to its pathophysiology]. *Brain and Nerve*, 61, 558–568.

Storyline in dreams through the night.

Trosman, H., et al. (1960). Studies in psychophysiology of dreams. *Archives of General Psychiatry*, 3, 602–607. See page 604.

Comparison between REM and NREM dream content.

For a review of recent studies, see McNamara, P., et al. (2010). REM and NREM sleep mentation. *International Review of Neurobiology*, 92, 69–86.

Cicogna, P., et al. (2000). Slow wave and REM sleep mentation. *Sleep Research Online: SRO*, 3, 67–72.

Foulkes, D. (1964) Theories of dream formation and recent studies of sleep consciousness. *Psychological Bulletin*, 62, 236–247. See page 241.

Changes in REM dreams as the night progresses.

Cipolli, C., et al. (2015). Time-of-night variations in the story-like organization of dream experience developed during rapid-eye movement sleep. *Journal of Sleep Research*, 24, 234–240.

Comparisons between dream reports obtained in the laboratory in dream reports obtained at home.

For a review, see Domhoff, G. W. (2005). The content of dreams: Methodologic and theoretical implications. In M. H. Kryger, T. Roth, and W. C. Dement (Eds.), *Principles and Practices of Sleep Medicine* (4th Ed.), pp. 522–534. Philadelphia: W. B. Saunders.

Activation of areas of the brain during REM and REM sleep.

Many of these papers are quite technical. Here are some of the papers I relied upon.

Dang-Vu, T. T., et al. (2010). Functional neuroimaging insights into the physiology of human sleep. *Sleep*, 33, 1589–1603

Desseilles, M., et al. (2011). Cognitive and emotional processes during dreaming: A neuroimaging view. *Consciousness and Cognition*, 20, 998–1008.

De Gennaro, L., et al. (2011). Amygdala and hippocampus volumetry and diffusivity in relation to dreaming. *Human Brain Mapping*, 32, 1458–1470.

The characterization of "mysterious" about the posterior cingulate cortex is from Pearson, et al. (2011). Posterior cingulate cortex: Adapting behavior to a changing world. *Trends in Cognitive Science*, 15, 143–151.

Dream recall, importance of autobiographical memories.

Robbins, P. R., & Tanck, R. H. (1978). Early memories and dream recall. *Journal of Clinical Psychology*, 34, 729–731.

Dream recall, factors involved in, such as age, gender, mood, bizarreness of content, motivation, and personality differences.

Waterman, D., et al. (1993). Symposium: Dream research methodology: Methodological issues affecting the collection of dreams. *Journal of Sleep Research*, 2, 8–12.

Robbins, P. R., & Tanck, R. H. (1988). Interest in dreams and dream recall. *Perceptual and Motor Skills*, 66, 291–294.

Robbins, P. R., & Tanck, R. H. (1988–1989). Depressed mood, dream recall and contentless dreams. *Imagination, Cognition and Personality*, 8, 165–174.

Cipolli, C., et al. (1993). Bizarreness effect in dream recall. *Sleep*, 16, 163–170.

Blagrove, M., & Pace-Schott, E. F. (2010). Trait and neurobiological correlates of individual differences in dream recall and dream content. *International Review of Neurobiology*, 92, 155–180.

Dream recall, suggestions for improving.

See Robbins, P. R. (1988). *The Psychology of Dreams*, Chapter 16.

Dreams in animals.

I did not see the Disney film, but am relying on a paper by Jankovic, S. M., et al. (2006), translated from a Serbian Journal. [The first film presentation of REM sleep behavior disorder precedes the scientific debut by 35 years]. *Srpski Arhiv Za Celekupno Lekarstvo*, 134, 466–469.

Lewis, J. E. (2008). Dream reports of animal rights activists. *Dreaming*, 18, 181–200.

Dreaming in color.

Schwitzgebel, E. (2003). Do people still report dreaming in black and white? An attempt to replicate a questionnaire from 1942. *Perceptual and Motor Skills*, 96, 25–29.

Kahn, E., et al. (1962). Incidence of color in immediately recalled dreams. *Science*, 137, 1054–1055.

Okada, H., et al. (2011). Life span differences in color dreaming. *Dreaming*, 21, 213–220.

Emotions in dreams.

Hall, C. S., & Van de Castle, R. (1966). *The Content Analysis of Dreams*. New York: Appleton-Century-Crofts.

Sources of dreams.

Freud, S. (1955). *The Interpretation of Dreams*. New York: Basic Books.

Van Rijn, E., et al. (2015). The dream-lag effect: Selective processing of personally significant events during Rapid Eye Movement sleep, but not during Slow Wave Sleep. *Neurobiology of Learning and Memory*, 122, 98–109.

Nielsen, T. A., et al. (2004). Immediate and delayed incorporations of events into dreams: Further replication and implications for dream function. *Journal of Sleep Research*, 13, 327–336.

Malinowski, J. E., & Horton, C. L. (2014). Memory sources of dreams: The incorporation of autobiographical rather than episodic experiences. *Journal of Sleep Research*, 23, 441–447.

Fosse, M. J., et al. (2003). Dreaming and episodic memory: A functional dissociation? *Journal of Cognitive Neuroscience*, 15, 1–9.

Experimental stimuli, influence on dreaming.

Dement, W. (1958). *The physiology of dreaming*. Doctoral dissertation, University of Chicago.

Carpenter, K. A. (1988). The effects of positive and negative pre-sleep stimuli on dream experiences. *Journal of Psychology*, 122, 33–37.

Cipolli, C., et al. (2004). Incorporation of presleep stimuli into dream contents: Evidence for a consolidation effect on declarative knowledge during REM sleep? *Journal of Sleep Research*, 13, 317–326.

Important events and dreams (divorce, pregnancy, cancer).

A study of the dreams of women going through divorce was reported in a paper by Cartwright, R., et al. (2006). Relation of dreams to waking concerns. *Psychiatry Research*, 141, 261–270.

Blake, R. L., Jr., & Reimann, J. (1993). The pregnancy-related dreams of pregnant women. *The Journal of the American Board of Family Practice*, 6, 117–122.

Lara-Carrasco, J., et al. (2013). Maternal representations in the dreams of pregnant women: a prospective comparative study. *Frontiers in Psychology*, August 27; 4:551. doi: 10.3389/fpsyg. 2013.00551.

Wellisch, D. K., & Cohen, M. (2011). In the midnight hour: Cancer and nightmares. A review of theories and interventions in psycho–oncology. *Palliative & Supportive Care*, 9, 191–200.

Hunger, thirst and food dreams.

Most of the discussion for this topic was based upon sources (e.g., the views of Earnest Jones) included in an interesting paper written by Tore Nielsen and Russell A. Powell, entitled "Dreams of the Rarebit Fiend: Food and diet as instigators of bizarre and disturbing dreams." It was published in *Frontiers in Psychology*, 2015, 6, 47.

Blindness and dream content.

Bertolo, H., et al. (2003). Visual dream content, graphical representation and EEG alpha activity in congenitally blind subjects. *Brain Research. Cognitive Brain Research*, 15, 277 –284.

Kirtley, D., & Cannistraci, K. (1974). Dreams of the visually handicapped: Toward a normative approach. *American Foundation for the Blind, Research Bulletin*, 27, 111–133.

Typical dreams, frequency and lifetime prevalence.

Schredl, M.. & Piel, E. (2007). Prevalence of flying dreams. *Perceptual and Motor Skills*, 105, 657–660.

For a review of research, see Domhoff, G. W. (2005). The content of dreams: Methodologic and theoretical implications. In M. H. Kryger, T. Roth, and W. C, Dement (Eds.), *Principles and Practices of Sleep Medicine* (4th Ed.), pp. 522–534. Philadelphia: W. B. Saunders.

Schredl, M., et al. (2004).Typical dreams: Stability and gender differences. *Journal of Psychology*, 138, 485–494.

Yu, C. K-C. (2015). One hundred typical themes in most recent dreams, diary dreams, and dreams spontaneously recollected from last night. *Dreaming*, 25, 206–219.

Romantic dreams.

Mikulincer, M., et al. (2011). Individual differences in adult attachment are systematically related to dream narratives. *Attachment and Human Development*, 13, 105–123.

Selterman, D., et al. (2012). Script-like attachment representations in dreams containing current romantic partners. *Attachment and Human Development*, 14, 501–515.

Lucid dreaming.

The reference for the review of methods of inducing lucid dreaming is Stumbrys,

T., et al. (2012). Induction of lucid dreams: A systematic review of evidence. *Consciousness and Cognition*, 21, 1456–1475.

The study of brain functioning during lucid dreams was Dresler, M., et al. (2012). Neural correlates of dream lucidity obtained from contrasting lucid versus non-lucid REM sleep: A combined EEG/fMRI case study. *Sleep*, 35, 1017–1020.

Saunders, G. T., et al. (2016). Lucid dreaming incidents: A quality effects meta-analysis of 50 years of research. *Consciousness and Cognition*, 43, 197–215.

Schredl, M., & Erlacher, D. (2011). Frequency of lucid dreaming in a representative German sample. *Perceptual and Motor skills*, 112, 104–108.

Sexuality in dreams.

There are two major studies I have relied on for estimates of the frequency with which men and women report sexual activities or ideas in their dreams. The first is the pioneering study carried out by C. S. Hall and R. L. Van de Castle reported in *The Content Analysis of Dreams*. (1966). New York: Appleton-Century-Croft. The second study, carried out in Canada, was reported by A. Zadra. (2007). My reference was a note, "Sexual activity reported in dreams of men and women," reported in *Science Daily*, June 15, 2007.

Sexual symbolism in dreams: Two empirical studies.

Robbins, P. R., & Tanck, R. H. (1980). Sexual gratification and sexual symbolism in dreams: Some support for Freud's theory. *Bulletin of the Menninger Clinic*, 44, 49–58.

Robbins, P. R., et al. (1985). Anxiety and dream symbolism. *Journal of Personality*, 5, 17–22.

The hypothesis relating anxiety to symbolism was advanced in Wallach, M. A. (1960). Two correlates of symbolic sexual arousal: Level of anxiety and liking for aesthetic material. *Journal of Abnormal and Social Psychology*, 61, 396–401.

Freud's theory of dreams.

My primary source for Freud's theories about dreams is his seminal work *The Interpretation of Dreams*. The copy I have used over the years is Freud, S. (1955). *The Interpretation of Dreams*. New York: Basic Books.

My primary source for biographical information about Freud is the multivolume work by Ernest Jones. Jones, E. (1953). *The Life and Work of Sigmund Freud*. New York: Basic Books.

I have also drawn occasionally on the four-volume set *The Collected Papers of Sigmund Freud*, using the 1953 edition, published by the Hogarth Press.

Kohler, T., & Borchers, H. (1996). [Experimental psychological evaluation of Freud's dream theory. Replication and extension of an earlier study]. *Psychotherapie, Psychosomatik Mediizinische, Psychologie*, 46, 419–422.

Jung's views about dreams.

Jung, C. G. (1974). *Dreams*. Princeton: Princeton University Press.

Some modern theories about dreams.

Robbins, P. R., & Tanck, R. H. (1990). Theories of dreams held by American college students. *Journal of Social Psychology*, 131, 143–145.

Revonsuo, A. (2000). The interpretation of dreams: An evolutionary hypothesis of the function of dreaming. *The Behavioral and Brain Sciences*, 23, 877–901. The quotations concerning the evolutionary theory are from the abstract of this paper.

The description of the activation-synthesis hypothesis was taken from the abstract of Hobson, J. A., & McCarley, R. W. (1977). The brain as a dream state generator: An activation-synthesis hypothesis of the dream process. *American Journal of Psychiatry*, 134, 1335–1348.

Hobson, J. A., et al. (2000). Dreaming and the brain: Toward a cognitive neuroscience of conscious states. *The Behavioral and Brain Sciences*, 23, 793–842.

Schredl, M., & Hofmann, F. (2003). Continuity between waking activities and dream activities. *Consciousness and Cognition*, 12, 298–308.

Foulkes, D., & Domhoff, G. W. (2014). Bottom-up or top-down in dream neuroscience? A top-down critique of two bottom-up studies. *Consciousness and Cognition*, 168–171.

Recurrent dreams.

Robbins, P. R., & Houshi, F. (1983). Some observations on recurrent dreams. *Bulletin of the Menninger Clinic*, 47, 262–265.

Robbins, P. R., & Tanck, R. H. (1991–1992). A comparison of recurrent dreams reported from childhood and recent recurrent dreams. *Imagination, Cognition and Personality*, 11, 259–262.

Gauchat, A., et al. (2015). The content of recurring dreams in young adolescents. *Consciousness and Cognition*, 37, 103–111.

Gender differences in dream content.

Numerous studies have shown that women have higher levels of dream recall then men. This difference has been confirmed through meta-analysis. For studies showing the tendency of women to more often share their dreams with other people, see Robbins, P. R., & Tanck, R. H. (1988). Theories of dreams held by American college students. *Journal of Social Psychology*, 131, 143–145 and Schredl, M. (2009). Sharing dreams: Sex and other sociodemographic variables. *Perceptual and Motor Skills*, 109, 235–238.

Hall, C.S., & Van de Castle, R. (1966). *The Content Analysis of Dreams*. New York: Appleton-Century-Crofts.

Zadra, A. (2007). Sexual activity reported in dreams of men and women. *Science Daily*, 15 June.

Domhoff, G. W. (2005). The content of dreams: Methodologic and theoretical implications. In M. H. Kryger, T. Roth, and W. C, Dement (Eds.), *Principles and Practices of Sleep Medicine* (4th Ed.), pp. 522–534. Philadelphia: W. B. Saunders.

Rubinstein, K., & Krippner, S. (1991). Gender differences and geographical differences in content from dreams elicited by a television announcement. *International Journal of Psychosomatics*, 38, 40–44.

Dreams of children.

Sandor, P., et al. (2015). Content analysis of 4- to 8-year-old children's dream reports. *Frontieres in Psychology*, April 30:6:534. doi. 10.3389/f psyg.2015.00534.

Foulkes, D. (1999). *Children's Dreaming and the Development of Consciousness*. Cambridge: Harvard University Press.

Muris, P., et al. (2000). Fears, worries, and scary dreams in 4- to 12-year-old children: Their content, developmental pattern, and origins. *Journal of Clinical Child Psychology*, 29, 43–52.

Dreams of the elderly.

Dale, A., et al. (2015). Ontogenetic patterns in the dreams of women across the lifespan. *Consciousness and Cognition*, 37, 214–224.

Grant, P., et al. (2014). The significance of end-of-life dreams and visions. *Nursing Times*, 110, 22–24.

Blick, K. A., & Howe, J. B. (1984). A comparison of the emotional content of dreams recalled by young and elderly women. *Journal of Psychology*, 116, 143–146.

Waterman, D. (1991). Aging and memory for dreams. *Perceptual and Motor Skills*, 73, 355–365.

Brenneis, C. B. (1975). Developmental aspects of aging in women: A comparative study of dreams. *Archives of General Psychiatry*, 32, 429–435.

Bizarre dreams.

The bizarre dream illustration was taken from the paper by Scarone, et al. See Scarone, S., et al. (2008). The dream as a model for psychosis: An experimental approach using bizarreness as a cognitive marker. *Schizophrenia Bulletin*, 34, 515–522. The brief citation from Jung was also taken from this paper.

Hobson, J. A., et al. (1987). Dream bizarreness and the activation-synthesis hypothesis. *Human Neurobiology*, 6, 157–164.

Malinowski, J. E., et al. (2014). The effect of time of night on wake-dream continuity. *Dreaming*, 24, 253–269.

Dreams of schizophrenics.

Gottesmann, C. (2006). Dreaming and schizophrenia: A common neurobiological background? *Medical Science*, 22, 201–205.

Carrington, P. (1972). Dreams and schizophrenia. *Archives of General Psychiatry*, 26, 343–350.

Limosani, L., et al. (2011). Bizarreness in dream reports and waking fantasies of psychotic schizophrenic and manic patients: Empirical evidences and theoretical consequences. *Psychiatry Research*, 189,195–199.

Dreams of depressed people.

Schredi, M. (1995). [Dream recall in depressed patients]. *Psychotherapie, Psychosomatk Medizinische, Psychologie*, 45, 414–417.

Luca, A., et al. (2013). Sleep disorders and depression: Brief review of the literature, case report, and nonpharmacologic interventions for depression. *Clinical Interventions in Aging*, 8, 1033–1039.

Robbins, P. R., & Tanck, R. H. (1988–1989). Depressed mood, dream recall and contentless dreams. *Imagination, Cognition and Personality*, 8, 165–174.

The illustration of the dream in the Venice coffee shop was taken from Miller, J. B. (1969). Dreams during various stages of depression. *Archives of General Psychiatry*, 20, 560–565.

McNamara, P., et al. (2010). Impact of REM sleep on distortions of self-concept, mood and memory in depressed/anxious participants. *Journal of Affective Disorders*, 122, 198–207.

Tribi, G. G., et al. (2013). Dreaming under antidepressants: A systematic review on evidence in depressive patients and healthy volunteers. *Sleep Medicine Review*, 17, 133–142.

Quartini, A. (2014). Changes in dream experience in relation with antidepressant escitalopram treatment in depressed female patients: A preliminary study. *Rivista Psichiatria*, 49, 187–191.

Dreams of anxious people.

Mindel, J. A., & Barrett, K. M. (2002). Nightmares and anxiety in elementary-aged children: Is there a relationship. *Child: Care, Health and Development*, 28, 317–322.

Nielsen, T. A., et al. (2000). Development of disturbing dreams during adolescence and their relation to anxiety symptoms. *Sleep*, 23, 727–736.

Gentil, M. L., & Lader, M. (1978). Dream content and daytime attitude in anxious and calm women. *Psychological Medicine*, 8, 297–304.

Nightmares in adults.

Robert, G., & Zadra, A. (2014). Thematic and content analysis of idiopathic nightmares and bad dreams. *Sleep*, 37, 409–417.

Paul, F., et al. (2015). Nightmares affect the experience of sleep quality but not sleep architecture: An ambulatory polysomnographic study. *Borderline Personality Disorder and Emotion Dysregulation*, 2, 3.

Levin, R., & Fireman, G. (2002). Nightmare prevalence, nightmare distress, and self-reported psychological disturbance. *Sleep*, 25, 205–212.

Nightmares in children.

Mindell, J. A., & Barrett, K. M. (2002). Nightmares and anxiety in elementary aged-children: Is there a relationship. *Child: Care, Health and Development*, 28, 317–322.

Gauchat, A., et al. (2014). Prevalence and correlates of disturbed dreaming in children. *Pathologie-Biologie*, 62, 311–318.

Wolke, D., & Lereya, S. T. (2014). Bullying and parasomnias: A longitudinal cohort study. *Pediatrics*, 134, 1040–1048.

Night terrors versus nightmares.

Sleep terrors—overview and facts. Sleep Education. A sleep health information resource by the American Academy of Sleep Medicine. http:www.sleepeducation.org/sleep-disorders-by-category/parasmonias/sleep-terrors/. For an estimate of the prevalence of sleep terrors, see Biorvatn, B., et al. (2010). Prevalence of different parasomnias in the general population. *Sleep Medicine*, 11, 1031–1034.

Guzman, C. S., & Wang, Y. P. (2008). Sleep terror disorder: A case report. *Revista Brasileira de Psiquiatria*, 30, No. 2.

Treatment of nightmares.

The Mayo Clinic's review of the treatment of nightmares with Prazosin was cautiously worded, stating that it was a well-tolerated medication that had a small but positive evidence base for treating PTSD-associated nightmares. Kung, S., et al. (2012). Treatment of nightmares with Prazosin: A systematic review. *Mayo Clinic Proceedings*, 87, 890–900.

The Best Practice Guide for the Treatment of Nightmare Disorder in Adults also recommended Prazosin for the treatment of posttraumatic stress disorder PTSD-associated nightmares. Its recommendation was less guarded. Aurora, R. N., et al. (2010). Best practice guide for the treatment of nightmare disorder in adults. *Journal of Clinical Sleep Medicine*, 6, 389–401.

Berlin, K. L., et al. (2010). Nightmare reduction in a Vietnam veteran using imagery rehearsal therapy. *Journal of Clinical Sleep Medicine*, 6, 487–488.

Seda, G., et al. (2015). Comparative meta-analysis of Prazosin and imagery rehearsal therapy for nightmare frequency, sleep quality, and posttraumatic stress. *Journal of Clinical Sleep Medicine*, 11, 11–22.

Krakow, B., & Zadra, A. (2006). Clinical management of chronic nightmares: Imagery rehearsal therapy. *Behavioral Sleep Medicine*, 4, 45–70.

Some of the ideas for helping children who experience nightmares were based on suggestions offered in Nightmares for Children, How to Help Kids with Nightmares. Cleveland Clinic Children's Diseases and Conditions. http://my. clevelandclinic.org/children's-hospital.

Positive effects from dreaming (mood, problem solving and creativity).

Freud's view of the positive benefits of interpreting anxiety dream was taken from Chapter IV (Distortion in Dreams), page 135, of Freud, S. (1955). *The Interpretation of Dreams*. New York: Basic Books.

Van der Helm, E., et al. (2011). REM sleep depotentiates amygdala activity to previous emotional experiences. *Current Biology*, 21, 2029–2032.

Beijamini, F., et al. (2014). After being challenged by a video game problem. Sleep increases the chance to solve it. *PLoS One*, January 9(1):e84342. doi: 10.1371/Journal.pone.008-4342.

Wamsley, E. J., et al. (2010). Dreaming of a learning task is associated with enhanced sleep-dependent memory consolidation. *Current Biology* 20, 850–855.

For references on the effects of sleep on memory consolidation, see both Beijamini, et al. and Wamsley, et al.

Wagner, U., et al. (2004). Sleep inspires insight. *Nature*, 427, 352–355.

Loewi's account of his dream was excerpted from Cai, D, J., et al. (2009). REM, not incubation, improves creativity by priming associative networks. *Proceedings National Academy of Sciences*, 106, 10130–10134.

Barrett, D. (1993). The "committee of sleep": A study of dream incubation for problem solving. *Dreaming*, 3, 115–122.

The description of Poincaré's experience was excerpted from Ritter, S. M., & Dijksterhuis, A. P. (2014). Creativity—the unconscious foundations of the incubation period. *Frontiers in Human Neuroscience*, 8, 215.

The dream experience of Coleridge in writing Kubla Kahn was cited in Thomas, W., and Brown, S. G. (1941). *Reading Poems: An Introduction to Critical Study*. New York: Oxford University Press.

Horan, N. (2014). *Under the Wide and Starry Sky*. New York: Random House.

Message and predictive dreams.

Oppenheim, A. L. (1956). The interpretation of dreams in the Ancient Near East with a translation of an Assyrian Dream—Book. *Transactions of the American Philosophical Society*, 46, Part 3, 179–371.

Ullman, M. (1969). Telepathy and Dreams. *Experimental Medicine and Surgery*, 27, 19–38.

Shainberg, D. (1976). Telepathy in psychoanalysis: An instance. *American Journal of Psychotherapy*, 30, 463–472.

Herodotus. (1942). *The Persian Wars*. New York: The Modern Library, Random House.

Sophocles. (1942). *Oedipus the King. The Complete Greek Tragedies*. Volume II. Chicago: University of Chicago Press.

Freud, S. (1953). The Occult Significance of Dreams. In J. Strachey (Ed.), *Collected Papers*, 5, *The International Psycho-Analytical Library*, No. 37, 158–162. The discussion is on page 159. London: Hogarth Press.

Part II: What Dreams May Tell Us About Ourselves

Hall, C. S., & Van de Castle, R. (1966). *The Content Analysis of Dreams*. New York: Appleton-Century-Crofts.

See in particular, Freud, S. (1955). *The Interpretation of Dreams*, pp. 100–104. New York: Basic Books.

Cavallero, C. (1987). Dream sources, associative mechanisms, and temporal dimension. *Sleep*, 10, 78–83.

A classic illustration of the use of Freud's technique is the dream of Irma. See Freud, S. (1955). *The Interpretation of Dreams*, pp. 106–121. New York: Basic Books.

A number of psychoanalytic writers have offered commentary about Freud's dream of Irma. An example would be Mautner, B. (1991). Freud's Irma dream: A psychoanalytic interpretation. *International Journal of Psychoanalysis*, 72 (Pt. 2). 275–286. Mautner hypothesized that the dream about Irma was really a cover for a repressed incident that haunted Freud—an incident of erotic aggression carried out by Freud against his sister Anna when he was five years old and she was three years old. This kind of analysis reinforces my view that psychoanalytic writers tend to engage in theoretical speculations rather than carry out scientific research to verify their theories. In any event, I react to such analyses with my usual attitude of skepticism.

Freud's concept of the dream work and the use of such mechanisms as condensation, displacement and symbols.

Discussions of the dream-work and the mechanisms it uses are considered in many places in Freud's *The Interpretation of Dreams*. Freud's belief that the dream-work combines the various sources of the dream into a single entity is stated on p. 179 of that work.

Critique of Freud's theory.

Kohler, T., & Borchers, H. (1996). [Experimental psychological evaluation of Freud's dream theory. Replication and extension of an earlier study]. *Psychotherapie Psychosomatik Medizinische Psychologie*, 12, 419–422.

Andresen, J. J. (1980). Rapunzel: The symbolism of the cutting of hair. *Journal of the American Psychoanalytic Association*, 28, 69–88. The quotation about the standards of evidence in psychoanalysis was taken from the abstract of the paper.

Approach of Carl Gustav Jung to dream analysis.

Jung, C. G. (1974). *Dreams*. Princeton: Princeton University Press. The quotations from Jung's writing about dream analysis were taken from pp. 99–101 and 104–105.

Zhu, C. (2013). Jung on the nature and interpretation of dreams: A developmental delineation with cognitive neuroscientific responses. *Behavioral Sciences* (Basil) 3, 662–675.

For an attempt at reconciliation between brain studies and Jung's theory, see Wilkinson, M. (2006). The dreaming mind-brain: A Jungian perspective. *Journal of Analytical Psychology*, 51, 43–59.

Kluger, H. Y. (1975) Archetypal dreams and "everyday" dreams: A statistical investigation into Jung's theory of the collective unconscious. *Israeli Annals of Psychiatry and Related Disciplines*, 13, 6–47.

Domino, G. (1976). Compensatory aspects of dreams: An empirical test of Jung's theory. *Journal of Personality and Social Psychology*, 34, 658–662.

Dream dictionaries.

Stratos, A. (2015). Egypt, perchance to dream: Dreams and their meaning in ancient Egypt. http://www.touregy pt.net/featurestories.htm.

Robbins, P. (2010). *Red Spotted Ox: A Pokot life*. IWGIA Document 124. Copenhagen, Denmark.

Oppenheim, L. (1956). The interpretation of dreams in the Ancient Near East with a translation of an Assyrian Dream—Book. *Transactions of the American Philosophical Society*, 46, part 3, 179–371.

Weiss, H. B. (1994). Oneirocritica Americana. *Bulletin of the New York Public Library*, 48, 519–541.

The three references used in the discussion of dream dictionaries are:

Dream book A. Fontana, D. (1997). *Teach Yourself to Dream: A Practical Guide to Unleashing the Power of the Subconscious Mind*. San Francisco: Chronicle Books.

Dream book B. Crisp, T. (2002). *Dream Dictionary: An A to Z Guide to Understanding your Unconscious Mind*. New York: Dell.

Dream book C. Lennox, M. (2015). *Llewellyn's Complete Dictionary of Dreams: Over 1,000 Dream Symbols and their Universal Meanings*. Woodbury, MN: Llewellyn Publications.

Coen, S. J., & Bradlow, P. A. (1985). The common mirror dream, dreamer, and the dream mirror. *Journal of the American Psychoanalytic Association*, 33, 797–820.

A widely used psychological test measuring narcissism is the Narcissistic Personality Inventory. See for example Raskin, R., & Terry, H. (1988). A principal-components analysis of the Narcissistic Personality Inventory and further evidence of its construct validity. *Journal of Personality and Social Psychology*, 54, 890–902.

Dream Incident Technique.

Robbins, P. R. (1988). The first three cases studies were taken from my book, *The Psychology of Dreams*. Jefferson, NC: McFarland.

Robbins, P. R. (1966). An approach to measuring psychological tension by means of Dream Associations. *Psychological Reports*, 18, 959–971.

Robbins, P. R.. & Tanck, R. H. (1978). The dream incident technique as a measure of unresolved problems. *Journal of Personality Assessment*, 42, 583–591.

Robbins, P. R., et al. (1972). A study of three psychosomatic hypotheses. *Journal of Psychosomatic Research*, 16, 93–97.

Tanck, R. H., & Robbins, P. R. (1970). Pupillary reactions to sexual, aggressive and other stimuli as a function of personality. *Journal of Projective Techniques and Personality Assessment*, 34, 277–282.

Robbins, P. R., & Tanck, R. H. (1969). Community violence and aggression in dreams: An observation. *Perceptual and Motor Skills*, 29, 41–42.

The study of pregnant women was carried out by F. Borgeat, et al. (1994). Perceptual defense and vulnerability to postpartum depression. *Acta Psychiatrica Scandinavica*, 90, 455–458.

Thirty-five items used for rating dream related incidents.

1. To lead or direct the activity.
2. To be daring, to seek adventure.
3. To protect, care for.
4. To excel: to attain a high standard.
5. To have my ideas, my way of doing things prevail.
6. To behave in such a way that I would not feel guilty about it afterwards.
7. To feel close to someone of the opposite sex.
8. To resist attempts by others to tell me what to do.
9. To behave in a way that I would later feel was completely responsible.
10. To physically attack; to hit, to strike, to slap.
11. To be the center of attention.
12. To find amusement; to have fun.
13. To avoid embarrassment; to avoid looking foolish or inadequate.
14. To fulfill an obligation; to live up to a commitment or promise.
15. To go to new places; to do new things.
16. To experience sexual pleasure.
17. To express anger; to tell someone off.
18. To have a good time; to enjoy myself.
19. To avoid hurting other persons.
20. To do something my conscience told me I must do.
21. To receive love or affection.
22. To avoid trouble or unpleasantness.
20. To give comfort, sympathy.
24. To get into the limelight.
25. To be independent; to be free to do things my own way.
26. To keep other people from realizing how nervous I felt.
27. To help someone in need or trouble.
28. To do something unusual and exciting.
29. To do something better than other people.
30. To share an experience with a friend.
31. To injure and inflict pain; to hurt someone.
32. To kiss, to pet—to make love.
33. To get the situation over with as quickly as possible.
34. To attract attention; to be noticed, talked about.
35. To accomplish, strive, achieve.

Twelve motivations measured by the DIT with identification of the specific items used to assess the motivation.

Affection (seeking warmth and love from other people) 7, 21, 30
Achievement (the wish to accomplish and excel) 4, 29, 35
Dominance (the wish to control others) 1, 5
Autonomy (to behave as one wishes and be from control) 8, 25
Adventure (to seek new experiences) 2, 15, 28
Sex (to engage in sexual activities) 16, 32
Aggression (feeling angry, wanting to hurt or retaliate) 10, 17, 31
Social Recognition (receiving the attention and admiration of others) 11, 24, 34
Nurturance (giving emotional support and comfort to others) 3, 19, 23, 27
Play (wanting recreation, a good time) 12, 18
Avoidance of embarrassment, unpleasantness (Murray called this "infavoidance," a term rarely used) 13, 22, 26, 33
Avoidance of guilt (Doing the responsible thing) 6, 9, 14, 20

Index